International Schools for Computer Scientists

Series Editors

ALFREDO FERRO

Department of Mathematics
University of Catania
Viale A. Doria 6
I-95125 Catania
Italy

DANIELE MUNDICI

Department of Computer Science
University of Milan
Via Comelico 39-41
I-20135 Milan
Italy

Advisory Editor

DOV M. GABBAY

Department of Computing
Imperial College
180 Queen's Gate
London SW7 2BZ
UK

International Schools for Computer Scientists

Giorgio Levi (ed.): *Advances in logic programming theory*
Egon Börger (ed.): *Specification and validation methods*

Advances in Logic Programming Theory

Edited by

GIORGIO LEVI

University of Pisa, Italy

CLARENDON PRESS · OXFORD

1994

Oxford University Press, Walton Street, Oxford OX2 6DP
Oxford New York
Athens Auckland Bangkok Bombay
Calcutta Cape Town Dar es Salaam Delhi
Florence Hong Kong Istanbul Karachi
Kuala Lumpur Madras Madrid Melbourne
Mexico City Nairobi Paris Singapore
Taipei Tokyo Toronto
and associated companies in
Berlin Ibadan

Oxford is a trade mark of Oxford University Press

Published in the United States by
Oxford University Press Inc., New York

© The contributors listed on pp. xiii–xiv, 1994

All rights reserved. No part of this publication may be
reproduced, stored in a retrieval system, or transmitted, in any
form or by any means, without the prior permission in writing of Oxford
University Press. Within the UK, exceptions are allowed in respect of any
fair dealing for the purpose of research or private study, or criticism or
review, as permitted under the Copyright, Designs and Patents Act, 1988, or
in the case of reprographic reproduction in accordance with the terms of
licences issued by the Copyright Licensing Agency. Enquiries concerning
reproduction outside those terms and in other countries should be sent to
the Rights Department, Oxford University Press, at the address above.

This book is sold subject to the condition that it shall not,
by way of trade or otherwise, be lent, re-sold, hired out, or otherwise
circulated without the publisher's prior consent in any form of binding
or cover other than that in which it is published and without a similar
condition including this condition being imposed
on the subsequent purchaser.

A catalogue record for this book is available from the British Library

Library of Congress Cataloging in Publication Data
(Data available)
ISBN 0 19 853853 7

Typeset by the editor and contributors on T_EX

Printed in Great Britain by
Bookcraft (Bath) Ltd
Midsomer Norton, Avon

Preface

Logic programming has emerged in the last fifteen years as one of the most promising new programming paradigms and as a very active research area. The area was started by two relevant events of the late seventies, i.e.

- the design of the programming language PROLOG, which soon became one of the most popular tools for declarative programming, and
- the realization that a subset of first order logic had a very natural interpretation as a programming language, which set the foundations of the theory of logic programming.

On one hand, the PROLOG experience has shown that several relevant problems, in areas such as expert systems, deductive data bases, knowledge representation and rapid prototyping, can profitably be tackled by means of the logic programming technology, by exploiting its unique features, which include unification, nondeterministic search and metaprogramming. On the other hand, it has shown that, by means of sophisticated optimization and implementation techniques, the performance of PROLOG systems can be made comparable to that of more traditional programming languages. The concern for the issue of parallelism has affected the development of new parallel implementations and the design of a new class of languages, the concurrent logic languages.

On the theory side, the main feature of logic programming is its simple semantics, directly based on the proof and model theories of first order logic. This led to a successful specialization to the case of logic programs of general formal (semantics-based) methods, such as proving program properties, analysis (e.g. abstract interpretation and termination) and transformation (e.g. partial evaluation). A lot of the recent achievements in the theory of logic programs are somewhat related to extensions of the basic positive logic language and to the related semantic problems. The main issue here is the semantics of negation in normal logic programs, where the original non-monotonic negation as failure rule has been extended in various ways and provided with new declarative characterizations. Other relevant new language constructs are

- constraints, which lead to a very important extension of the paradigm which allows us to compute on new domains,
- concurrency,
- modules and objects.

The classical theory of logic programming, i.e. the semantics of positive logic programs and the theory of negation as failure-completion, are described in several good textbooks for postgraduate courses. This volume can be a useful integration, which covers, for the first time in a systematic way, some additional relevant topics.

The chapter by John Schlipf considers various approaches to the problem of modeling negation as failure by means of semantics capturing common sense reasoning. The paper discusses a systematic unified framework which can be used to compare various semantics proposed in the literature.

The chapter by Frank de Boer and Catuscia Palamidessi considers the field of concurrent logic programming and presents the motivations, the principal lines along which the field has developed, the various paradigms which have been proposed, and the main approaches to the semantic foundations.

The chapter by Saumya Debray gives a broad overview of the field of dataflow analysis of logic programs. Starting with a general discussion of abstract interpretation, a formal framework that underlies much of the work on static program analysis, it discusses some of the wide variety of frameworks that have been proposed for the analysis of logic programs, considers the relationship between complexity and precision of analysis, and examines some specific analyses that have been proposed in the literature.

The chapter by Krzysztof Apt and Dino Pedreschi provides a uniform and simplified presentation of the methods of Bezem and of Apt and Pedreschi for proving termination of logic and PROLOG programs and shows how these methods can be refined so that they can be used in a modular way.

Finally, the chapter by Egon Börger and Elvinia Riccobene shows an example of application of Gurevich's notion of evolving algebras to the specification of the semantics of control. The case which is considered is the language Gödel, which is a fully declarative successor of PROLOG.

The papers included in this collection are based on the notes of some of the lectures delivered at the Fourth International School for Computer Science Researchers on *Foundations of Logic Programming*. The school was held in Acireale, Italy, from June 22 to July 4 1992 and was organized by Alfredo Ferro from the Università di Catania. I gratefully acknowledge

- Alfredo Ferro,
- all the authors of the articles collected in this volume,
- Michael J. Maher and Dale Miller, who delivered excellent lectures

in Acireale,
- the Scientific Committee of the School for Computer Science Researchers,

for their contribution in making the school such a successful event.

Giorgio Levi
Pisa, January 1994

Contents

1 A Comparison of Notions of Negation as Failure   1
John S. Schlipf
1 Introduction   1
2 Definite programs   2
3 Normal logic programs   5
4 Intuitions for negation as failure   6
5 Some standard examples   8
6 A semantics survey   12
   6.1 Classical logic   12
   6.2 Minimal model semantics   12
   6.3 Perfect models of stratified programs   14
   6.4 2-valued program completion semantics   19
   6.5 3-valued program completion semantics   22
   6.6 The stable and well-founded semantics   27
   6.7 Summary: a comparison of inferences made   35
   6.8 Other approaches   36
7 Intuitions   36
   7.1 On precedences and the minimal model semantics   36
   7.2 Intuitions: 3-valued and 2-valued   37
8 Formalisms for comparing semantics   40
   8.1 Standards of acceptability as a logic   41
   8.2 Standards of acceptability for programming   43
9 Some comments on expressive powers   47
Bibliography   50

2 From Concurrent Logic Programming to Concurrent Constraint Programming   55
Frank S. de Boer and Catuscia Palamidessi
1 Introduction   55
   1.1 Don't know versus don't care nondeterminism   56
   1.2 Synchronization and communication mechanisms   56
   1.3 The influence of Constraint Logic Programming   57
   1.4 Concurrent constraint programming   57
   1.5 Semantics   58
   1.6 Current trends   60
   1.7 Plan of the paper   61
2 Preliminaries.   61
3 Concurrent logic languages   62

	3.1	Process interpretation of logic programming	63
	3.2	Explicit mechanisms for concurrency	67
	3.3	Variables as input output channels	76
	3.4	An alternative approach to synchronization: P-PROLOG	82
4	The ALPS languages		83
	4.1	Constraint systems	84
	4.2	The validation rule	85
5	Concurrent constraint programming		86
	5.1	The language	86
	5.2	Operational semantics	87
	5.3	Embedding concurrent logic programming	89
	5.4	Denotational semantics	92
	5.5	Reactive sequences versus closure operators	100
Bibliography			107

3 Formal Bases for Dataflow Analysis of Logic Programs 115
Saumya K. Debray

1	Introduction		115
2	Definitions		116
3	Abstract Interpretation		119
	3.1	Overview	119
	3.2	The Galois connection approach	122
	3.3	The widening/narrowing approach	127
	3.4	The hybrid approach	129
	3.5	Approximating the greatest fixpoint	130
4	Frameworks for dataflow analysis of logic programs		131
	4.1	Frameworks for top-down analyses	131
	4.2	Frameworks for bottom-up analyses	137
	4.3	Algebraic approaches	140
	4.4	"Magic-Set"-based approaches	141
5	Complexity and efficiency issues		143
	5.1	The relationship between complexity and precision	143
	5.2	Deriving efficient analysis algorithms	144
	5.3	Experimental results	151
6	Applications		152
	6.1	Type inference	152
	6.2	Groundness analysis	156
	6.3	Occur check optimization	158
	6.4	Compile-time garbage collection	159
	6.5	Argument size analysis	161
	6.6	Cost analysis	162
7	Analysis of concurrent logic programs		163

8	Analysis of Constraint Logic Programs	166
9	What information do implementors really want?	168
10	Inter-module analysis	170
Bibliography		171

4   Modular Termination Proofs for Logic and Pure PROLOG Programs   183
Krzysztof R. Apt and Dino Pedreschi

1	Introduction	183
	1.1 Motivation	183
	1.2 Preliminaries	184
2	Termination	185
	2.1 Motivation	185
	2.2 Terminating programs	186
	2.3 Examples	191
3	Left termination	192
	3.1 Motivation	192
	3.2 Left terminating programs	194
	3.3 Examples	200
4	A modular approach to termination	204
	4.1 Drawbacks of the proof method	204
	4.2 Semi-recurrent programs	206
	4.3 Methodology	210
	4.4 Examples	211
5	A modular approach to left termination	216
	5.1 Semi-acceptable programs	216
	5.2 Examples	218
Bibliography		228

5   Logic + Control Revisited: an Abstract Interpreter for GÖDEL Programs   231
Egon Börger and Elvinia Riccobene

1	Introduction	231
2	Evolving algebras	233
3	Signature of GÖDEL algebras	235
4	Delaying computations	238
5	Transition rules	239
	5.1 Initialization and rules for user-defined predicates	239
	5.2 The pruning operator	241
	5.3 The computation of negative literals	244
	5.4 The computation of conditionals	246
6	Depth-first search and concurrency	250
7	Conclusion and outlook	251
8	Acknowledgements	252

Bibliography 252

Contributors

Krzysztof R. Apt
CWI
P.O. Box 4079
1009 AB Amsterdam, The Netherlands
and
Faculty of Mathematics and Computer Science
University of Amsterdam
Plantage Muidergracht 24
1018 TV Amsterdam, The Netherlands
email: apt@cwi.nl

Egon Börger
Dipartimento di Informatica
Università di Pisa
Corso Italia 40
56125 Pisa, Italy
email: boerger@di.unipi.it

Frank S. de Boer
Department of Mathematics and Computer Science
Free University
de Boelelaan 1081
1081 HV Amsterdam, The Netherlands
email: frb@cwi.nl

Saumya K. Debray
Department of Computer Science
The University of Arizona
Tucson, AZ 85721, USA
email: debray@cs.arizona.edu

Catuscia Palamidessi
Dipartimento di Informatica e Scienze dell'Informazione (DISI)
Università di Genova
via Benedetto XV 3
16132 Genova, Italy
email: catuscia@di.unipi.it

Dino Pedreschi
Dipartimento di Informatica
Università di Pisa

Corso Italia 40
56125 Pisa, Italy email: pedre@di.unipi.it

Elvinia Riccobene
Dipartimento di Matematica
Università di Catania
Viale Andrea Doria 6
95125 Catania, Italy
email: riccobene@mathct.cineca.it

John S. Schlipf
Computer Science Department
University of Cincinnati
Cincinnati, OH 45221-0008, USA
email: john.schlipf@uc.edu

1
A Comparison of Notions of Negation as Failure

John S. Schlipf
University of Cincinnati

Abstract

When logic programming was generalized to allow negative subgoals, difficulties immediately arose concerning the meaning of negation as failure in the context of these subgoals. Various semantics have been proposed, each attempting to capture natural intuitions about negation as failure and to preserve other intuitive properties. We discuss several such semantics for normal logic programs here: the minimal model semantics, the perfect model semantics for stratified programs, the two- and three-valued program completion semantics, and the stable and well-founded semantics. We contrast them in various ways: we present some examples and intuitions that might be used to justify them, discuss which violate or preserve certain properties of more classical logics and which allow modular programming, and present some results about expressive powers and computational complexity.

1 Introduction

It is widely believed, at least in parts of the artificial intelligence community, that in many ordinary situations classical logic does not at all capture the common meaning of the word "not." The idea is that, in what McCarthy calls "common-sense reasoning," humans apply defaults for inferring information that cannot be proved. One common such default is called "negation as failure": essentially, if there is no reason to infer an atomic sentence to be true, infer it to be false. For example, if a physician sees no evidence that a patient has tuberculosis, then the physician normally infers that the patient does not have it. If there is no information in the University database to suggest that a certain person is a student at the University, then University officials infer that the person is not.

It is examples like these that have led researchers to try to add negation as failure as an inference rule in logic, notably in the context of logic programming. But there is also a much more practical reason to like negation as failure. Suppose one has an inference engine that can make inferences using negation as failure. There are circumstances where it is relatively easy to specify when a relation is true but much harder to specify directly when it is false. Consider the definition of the English word "ancestor": it is relatively easy to specify recursively when x is an ancestor of y, but there is no such straightforward recursive definition of "not an ancestor." In such cases, it seems desirable for a person using the inference engine to have to specify only when the relation is true, the easy part, and to leave the hard part for the automated system.

It turns out, however, that adding negation as failure is more complicated than it may at first appear. Consider the example

$$Bird(Tweety) \lor Bird(Opus).$$

This sentence does not imply *Bird(Tweety)*, so a simple-minded negation as failure would lead one to infer $\neg Bird(Tweety)$. Similarly, one would infer $\neg Bird(Opus)$. But inferring both of them contradicts the original assumption.

The "definition" of negation as failure given above is certainly vague. A goal of research in negation as failure semantics is to provide a semantics which captures common-sense reasoning, such as with the medical diagnosis and University database examples above, but avoids unnatural inferences, such as with the birds above.

This chapter compares various formalisms for negation as failure proposed for the class of *normal logic programs*.

2 Definite programs

A paradigmatic example of negation as failure comes from the van Emden Kowalski semantics for definite programs. A standard reference here is Lloyd's text (1987); we shall just state some basic theorems and define some basic vocabulary in this section.

Definition 2.1 *A definite rule is a universal sentence*

$$\forall X_1, \ldots, X_k(\alpha \leftarrow \beta_1, \beta_2, \ldots, \beta_n),$$

where $n \geq 0$, α and the β_i's are atomic formulas, and $X_1, \ldots, X_k$ is the set of variables appearing in α and the β_i's. A definite program is a (usually finite) set of definite clauses.

The commas in the rules above, specifically appearing on the right side of the $\leftarrow$ symbols, are treated as $\land$'s – "and"'s. The symbol $\leftarrow$ is taken as some sort of "if" type construct. In general, a rule is to be taken as some sort of inference rule: for any particular values of $X_1, \ldots, X_k$, if the β_i's are all true, then α must also be true. When we say that $\leftarrow$ is interpreted *classically*, we mean that it is treated as the usual (material) implication of classical logic, as the formula

$$\forall X_1, \ldots, X_k(\beta_1 \land \beta_2 \land \ldots \land \beta_n \rightarrow \alpha).$$

We shall consider ← to be a connective of formal logic. In particular, we shall consider a formula such as a definite rule

$$\forall X_1, \ldots, X_k (\alpha \leftarrow \beta_1, \beta_2, \ldots, \beta_n)$$

to be a universal sentence. We shall define later a normal logic program also to be a set of universal formulas.

Definite rules are also called *pure Horn clauses*. A definite rule

$$\forall X_1, \ldots, X_k (\alpha \leftarrow \beta_1, \beta_2, \ldots, \beta_n)$$

is usually written as just

$$\alpha \leftarrow \beta_1, \beta_2, \ldots, \beta_n$$

with the quantifiers implicit.

Example 2.2 The following is a definite program:

$$\begin{aligned}
even(0) &\leftarrow \\
even(s(X)) &\leftarrow odd(X) \\
odd(s(X)) &\leftarrow even(X) \\
exceptional(X) &\leftarrow even(X), odd(X).
\end{aligned}$$

Definition 2.3 *An atom, also called a positive literal, is a proposition letter a or an atomic formula $r(t_1, \ldots, t_k)$, where r is a k-ary relation symbol and the t_i's are terms. (We shall frequently treat proposition letters as 0-ary relation symbols.) A negative literal is a negated atom, $\neg r(t_1, \ldots, t_k)$. A literal is a positive or negative literal.*

Definition 2.4 *A term or formula is ground if it contains no variables. (Thus every proposition letter is a ground formula.) A ground instance of a universal sentence*

$$\psi = \forall X_1, \ldots, X_k \phi,$$

ϕ *quantifier-free, is a sentence*

$$\phi[X_1/t_1, \ldots, X_k/t_k]$$

formed from ϕ by replacing the variables $X_1, \ldots, X_k$ with ground terms $t_1, \ldots, t_k$. For $\mathcal{P}$ a set of universal formulas (such as a logic program), the ground instantiation of $\mathcal{P}$ is the set of all ground instances of sentences of $\mathcal{P}$; here it must be understood what the set of ground terms is: usually, it is the set of ground terms built up from the function and constant symbols appearing in $\mathcal{P}$.

Definition 2.5 *The Herbrand universe $\mathbf{U}_\mathcal{P}$ of a logic program $\mathcal{P}$ is the set of all ground terms of $\mathcal{P}$. The Herbrand base $\mathbf{B}_\mathcal{P}$ of $\mathcal{P}$ is the set of all ground positive literals of $\mathcal{P}$.*

A *preinterpretation* for a set of language of first order logic is an interpretation of the function and constant symbols of the language; thus it consists of a universe, an interpretation of each constant symbol as an element of the universe, and an interpretation of each function symbol as a function (of the same arity, i.e., number of arguments, as the function symbol) from the universe to the universe. The Herbrand preinterpretation of a logic program $\mathcal{P}$ is the interpretation with universe $\mathbf{U}_\mathcal{P}$ which interprets each ground term $t \in \mathbf{U}_\mathcal{P}$ as t itself. Following common practice in logic programming, we shall also call the Herbrand preinterpretation $\mathbf{U}_\mathcal{P}$ and refer to it as the *Herbrand universe*; context should determine whether we mean the actual universe or the preinterpretation.

An interpretation $\mathcal{M}$ for $\mathcal{P}$ is an *Herbrand interpretation* if it has the same universe and interpretations of the constant and function symbols as does $\mathbf{U}_\mathcal{P}$. An *Herbrand model* of $\mathcal{P}$ is an Herbrand interpretation of $\mathcal{P}$ which is a model of $\mathcal{P}$, where $\leftarrow$ is interpreted classically.

An Herbrand interpretation $\mathcal{M}$ for a program $\mathcal{P}$ is identified with the set of all formulas in $\mathbf{B}_\mathcal{P}$ true in $\mathcal{M}$. It is traditional in logic programming to restrict attention to Herbrand interpretations of logic programs.

Definition 2.6 *The van Emden Kowalski semantics for a definite program $\mathcal{P}$ is defined as follows:*

- *Let $\mathbf{T}_\mathcal{P}\uparrow 0 = \emptyset$.*
- *Let $\mathbf{T}_\mathcal{P}\uparrow n+1$ be the set of all ground literals α' such that there is a ground instance*
$$\alpha' \leftarrow \beta'_1, \ldots, \beta'_n$$
  of a rule of $\mathcal{P}$ where each $\beta'_i \in \mathbf{T}_\mathcal{P}\uparrow n$.
- *Let $\mathbf{T}_\mathcal{P}\uparrow\omega = \bigcup_i \mathbf{T}_\mathcal{P}\uparrow i$.*

The van Emden Kowalski semantics declares that $\mathbf{T}_\mathcal{P}\uparrow\omega$ (treated as an Herbrand interpretation of $\mathcal{P}$) is the intended interpretation of $\mathcal{P}$.

Thus, in particular, in the van Emden Kowalski semantics all ground atoms in $\mathbf{T}_\mathcal{P}\uparrow\omega$ are inferred to be true and all ground atoms in $\mathbf{B}_\mathcal{P} - \mathbf{T}_\mathcal{P}\uparrow\omega$ are inferred to be false. So, under this semantics, an atomic sentence is inferred to be false unless it can be proved true under the restrictive proof procedures captured in the construction of $\mathbf{T}_\mathcal{P}\uparrow\omega$.

Theorem 2.7

(1) $\mathbf{T}_\mathcal{P}\uparrow\omega$ *is an Herbrand model of $\mathcal{P}$.*
(2) *Every positive literal in $\mathbf{T}_\mathcal{P}\uparrow\omega$ is true in every model (not just every Herbrand model) of $\mathcal{P}$.*
(3) $\mathbf{T}_\mathcal{P}\uparrow\omega$ *is the intersection of all Herbrand models of $\mathcal{P}$ (considered as subsets of $\mathbf{B}_\mathcal{P}$).*

Thus the van Emden Kowalski semantics captures negation as failure in another way also: a ground atom is inferred to be false unless it is true in every model of the definite program.

The definition of the $\mathbf{T}_P \uparrow n$'s and the van Emden Kowalski semantics extends in an obvious way to the circumstance where the chosen universe is the Herbrand universe for some language larger than the language of the program. We shall use this fact later.

3 Normal logic programs

Research in logic programming semantics in the last few years has been concerned with developing suitable notions of negation as failure for larger classes of logic programs. In this chapter we shall discuss extensions to *normal logic programs*.

Definition 3.1 *A normal rule is a formula of the form*

$$\forall X_1, \ldots X_k(\alpha \leftarrow \beta_1, \beta_2, \ldots, \beta_n)$$

where α is a positive literal and the β_i's are literals, either positive or negative. The atom α is called the head of the rule, $\beta_1, \beta_2, \ldots, \beta_n$ is called the body, and the β_i's are called subgoals. More specifically, β_i is a positive (respectively, negative) subgoal if it is a positive (respectively, negative) literal.

The body of a rule is considered to be a set of subgoals. In particular, we freely reorder the subgoals, usually to put positive literals in front of negative literals, and eliminate duplicate subgoals.

A normal logic program is a (usually finite) set of normal rules.

A normal rule is also usually written without the universal quantifiers, with the quantifiers understood.

Some more recent research has concerned more general kinds of rules, notably rules with disjunctions of atoms in their heads and rules with negative atoms in their heads. But we shall not address such programs here.

Following the tradition in logic programming, we shall limit attention to Herbrand interpretations. Over Herbrand interpretations, a normal logic program (because it is a set of universal sentences) is equivalent to its ground instantiation. Moreover, since the ground instantiation is quantifier- and variable-free, the ground instantiation may be treated as a propositional logic program, with one proposition letter for each ground atomic formula (e.g., each element of the Herbrand base).

Henceforth, we shall generally define semantics of logic programs from the ground instantiations of those logic programs. Thus we shall essentially be defining the semantics directly only for propositional programs, defining the semantics of any other program to be the semantics of its ground instantiation. This makes the definitions simpler, though at a cost: the ground instantiation of a finite first order logic program containing function symbols will be infinite; it is for this reason that we have defined a logic program to be a *usually finite* set of rules, rather than always finite.

We declare at the outset that we are *not* concerned with any existing logic programming language, such as PROLOG, even though PROLOG exhibits some similarity to the semantics we are describing and at times might be thought of as partially motivating them. Beyond this:

- We are concerned only with *declarative* semantics: specifying what is to be inferred, but not specifying a computation rule (though we shall refer briefly to general questions of whether any suitable computation rules exist).
- We assume the semantics does *not* depend upon (1) the order of rules in a program, and (2) the order of subgoals in a program.
- We assume that logic programs contain no control information, such as *cut*.

4 Intuitions for negation as failure

With normal rules, there is no universally accepted semantics for negation as failure. Generalizing various properties of the van Emden Kowalski semantics for definite programs leads to differing semantics. We start out with a slightly more precise statement of negation as failure than we gave before:

Negation as failure: *If it is obvious (in some way) that an atomic statement cannot be proved (using some method of proof), infer the statement to be false.*

Depending upon how one interprets the "in some way" and "using some method of proof" above, one derives varying standards for negation as failure semantics. For example, one might well choose, as the method of proof, a proof system for classical logic. On the other hand, it is common in logic programming to consider much more restrictive methods of proof, often limiting attention to rather constructive proof rules. In particular, the $\mathbf{T}_\mathcal{P}$ construction given for the van Emden Kowalski semantics uses just three proof rules from classical logic:

$$\text{modus ponens:} \; \frac{\alpha \leftarrow \beta \quad \beta}{\alpha} \qquad \wedge\text{-introduction:} \; \frac{\alpha \quad \beta}{\alpha \wedge \beta}$$

$$\text{universal instantiation:} \; \frac{\forall X \alpha(X)}{\alpha(t)}$$

Of course, for normal logic programming applications, the list of logical inference rules can be limited partly because normal logic programs do not use, say, the ∃ and ∨ symbols of formal logic. But in addition the rules above do not allow, for example, proof by contradiction. It is common to limit the proof methods alluded to in the statement of negation as failure to some such limited list as above; in particular, proof by contradiction is frequently not allowed.

The divergence of definitions for negation as failure may also be due in part to the lack of common paradigmatic intuitions. Certainly in many circumstances negation as failure makes little sense. But there is a variety of intuitions leading to negation as failure, such as:

- Rules may be used implicitly as *definitions*.
- Rules may reflect *causal relationships*, and one may assume that no ground atom may be true without a cause.
- Rules may reflect *reasons to believe* atomic formulas. Normally, a rational person does not believe atomic sentences without reason. Note the ambiguity here between what is believed to be false and what is simply not believed to be true. In some cases it may be difficult to determine which is the proper notion.
- There are circumstances where negation as failure is tied in with the meaning of the word "or." "Or" can have two meanings: the inclusive $(a \vee b)$ and the exclusive $(a \underline{\vee} b)$; attempts to express exclusive "or" may lead to negation as failure.
- One might argue that classical logic captures much of what people mean, but that people *frequently tend* to use "if" in senses like the first three above. Thus negation as failure may be used as an *heuristic* to infer what a programmer *probably* had in mind.

As we present and discuss alternative definitions of negation as failure, the reader is invited to see which of the intuitions defined above seem to motivate each definition.

Various sorts of standards have been used for motivation and comparison of logic programming semantics. Some common ones are listed below. We shall return to such comparisons later in this chapter.

(1) **Justification by example:** This is the most common method: we find circumstances where a particular form of negation matches what seem to be normal intuitions.

(2) **Well-behavedness:** Various abstract properties of inference-drawing mechanisms have been proposed; these can sometimes be used as tools for evaluating the behavior of logic programming semantics. For example, with classical logic one has the properties below, where for any logic program $\mathcal{P}$, $Inferences(P)$ denotes the set of inferences the semantics makes from $\mathcal{P}$.

- **Monotonicity:** If $\mathcal{P} \subseteq \mathcal{Q}$, then

$Inferences(\mathcal{P}) \subseteq Inferences(\mathcal{Q})$.
- **Equivalence of a rule and its contrapositive:**
$Inferences(\{a \leftarrow \neg b\}) = Inferences(\{b \leftarrow \neg a\})$.

We shall reject monotonicity with negation as failure: an empty set of assumptions gives no reason to infer a, so we infer $\neg a$, but $\emptyset \subseteq \{a\}$, and from $\{a\}$ we certainly do not infer $\neg a$. Many versions of negation as failure also reject the equivalence of a rule and its contrapositive: the limited logical rules often used in logic programming do not permit proof of the equivalence of a statement and its contrapositive; this is often considered a desirable feature of logic programming semantics. Interestingly, somewhat different standards may be justified depending upon whether one is thinking of logic programming as a method of programming or as a method of describing, or specifying, reality.

(3) **Expressive power:** Let us say a set R of ground terms is *definable* in a semantics if, for some finite logic program $\mathcal{P}$ and some relation symbol r of $\mathcal{P}$,

$$R = \{t : r(t) \in Inferences(\mathcal{P})\}.$$

(There are other notions of definability by logic programs also used in the literature.) One can then compare semantics based upon what sets can be defined.

(4) **Computational difficulty:** Ease of computation is, of course, a desirable feature of any semantics. Unfortunately, there is a trade-off here: with the semantics we shall consider, difficulty of computation is closely tied to expressive power.

5 Some standard examples

We present here some examples of normal logic programs from the literature.

Example 5.1

(1) **Components in graphs:** Consider a graph given in a logic program by its list of edges:

$$edge(a, b) \leftarrow$$
$$edge(b, a) \leftarrow$$
$$\ldots$$

Then the following program *defines* the relationship of two vertices being in the same component:

$$sameComponent(X, X) \leftarrow$$
$$sameComponent(X, Z) \leftarrow sameComponent(X, Y), edge(Y, Z)$$

$$notSameComponent(X,Y) \leftarrow \neg sameComponent(X,Y)$$

(2) **Ancestry relationships:** Suppose a collection of family trees is given as a collection of facts, such as:

$$married(Henry, Catherine) \leftarrow$$
$$married(Catherine, Henry) \leftarrow$$
$$father(Henry, Mary) \leftarrow$$
$$mother(Catherine, Mary) \leftarrow$$
$$married(Henry, Anne) \leftarrow$$
$$married(Anne, Henry) \leftarrow$$
$$father(Henry, Elizabeth) \leftarrow$$
$$mother(Anne, Elizabeth) \leftarrow$$
$$\ldots$$

Based upon the relationships shown, the words "ancestor," "relative," and "inlaw" are defined by the program below. To simplify the program, we shall define each person to be an ancestor and a relative of him(her)self. An "inlaw" is a person who is a relative of a spouse, or a spouse of a relative, but not a relative him(her)self.

$$ancestor(X,X) \leftarrow$$
$$ancestor(X,Z) \leftarrow ancestor(X,Y), mother(Y,Z)$$
$$ancestor(X,Z) \leftarrow ancestor(X,Y), father(Y,Z)$$
$$relative(Y,Z) \leftarrow ancestor(X,Y), ancestor(X,Z)$$
$$notAncestor(X,Y) \leftarrow \neg ancestor(X,Y)$$
$$notRelative(X,Y) \leftarrow \neg relative(X,Y)$$
$$inlaw(X,Z) \leftarrow relative(X,Y), married(Y,Z),$$
$$notRelative(X,Z)$$
$$inlaw(X,Y) \leftarrow inlaw(Y,X)$$
$$notInlaw(X,Y) \leftarrow \neg inlaw(X,Y)$$

(3) **Winning positions in games:** Suppose two players take turns moving a peg from position to position in a game; the legal moves are given as a list of facts:

$$move(a,b) \leftarrow$$
$$move(a,c) \leftarrow$$
$$move(b,d) \leftarrow$$
$$\ldots$$

where the move relation can be illustrated with a graph such as below:

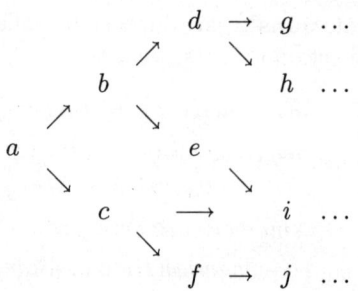

Suppose that the *move* relation is acyclic, and that the person who makes the last (legal) move in the game wins. This is defined by the following program:

$$winning(X) \leftarrow move(X,Y), \neg winning(Y)$$

(4) **Winning positions in games with cycles:** Repeat the above example, but let the *move* relationship have cycles. When does the notion of a *winning* position make sense? Does the semantics make reasonable inferences?

(5) **Expressing quantifiers** A *general rule* is a sentence of the form

$$\forall X_1, \ldots, X_k (a \leftarrow \phi),$$

where a is an atom, ϕ is an arbitrary first order formula, and where the free variables of a and ϕ are among $X_1, \ldots, X_k$. In some logic programming semantics it is possible to express general rules in terms of normal rules. The difficult part is to express a universal quantification of the body of a rule, such as
$$a(X) \leftarrow \forall Y (b(X,Y)).$$
First rephrase that as
$$a(X) \leftarrow \neg \exists Y (\neg b(X,Y))$$
and express that as two rules, using an extra predicate c,
$$c(X) \leftarrow \exists (Y)(\neg b(X,Y)), \text{ and}$$
$$a(X) \leftarrow \neg c(X).$$
Finally, rewrite that using the implicit quantification:
$$c(X) \leftarrow \neg b(X,Y) \text{ and}$$
$$a(X) \leftarrow \neg c(X).$$
In some of the semantics we study, this translation is correct. In those semantics, it turns out that it is also possible to express arbitrary general rules with normal rules. (Warning: some of the semantics we discuss here are based upon three-valued logics. For such logics, one must use the appropriate meanings of $\wedge, \vee, \neg, \forall$ and $\exists$. See Subsection 6.5.)

Example 5.2 The following propositional examples show various properties of the semantics we shall present.

(1) **Differences between rules and their contrapositives:**
$$\mathcal{P} = \{a \leftarrow \neg b\} \qquad \mathcal{Q} = \{b \leftarrow \neg a\}.$$

(2) **Positive recursive cycles:**
$$\mathcal{P} = \{p \leftarrow p\} \qquad \mathcal{Q} = \{a \leftarrow b, \ b \leftarrow a\}.$$

(3) **Negative recursive cycles, even length:**
$$\mathcal{P} = \{a \leftarrow \neg b, \ b \leftarrow \neg a\}$$
$$\mathcal{Q} = \{a \leftarrow \neg b, \ b \leftarrow \neg c, \ c \leftarrow \neg d, \ d \leftarrow \neg a\}.$$

(4) **Negative recursive cycles, odd length:**
$$\mathcal{P} = \{p \leftarrow \neg p\} \qquad \mathcal{Q} = \{p \leftarrow \neg q, \ q \leftarrow \neg r, \ r \leftarrow \neg q\}.$$

(5) **Factoring into cases and non-determinism**
$$\mathcal{P} = \{a \leftarrow \neg b, \ b \leftarrow \neg a, \ c \leftarrow a, \ c \leftarrow b\}$$
$$\mathcal{Q} = \{a \leftarrow \neg b, \ b \leftarrow \neg a, \ c \leftarrow \neg a, \ c \leftarrow \neg b\}.$$

(6) **Possible abductive inferences:** The first example, due to Van Gelder, consists of one even length negative recursion, one odd length negative recursion, and one additional rule.
$$\mathcal{P} = \{a \leftarrow \neg b, \ b \leftarrow \neg a, \ p \leftarrow \neg p, \ p \leftarrow \neg a\}.$$

The second example turns out to be rather similar.
$$\mathcal{Q} = \{a \leftarrow \neg b, \ b \leftarrow \neg a, \ c \leftarrow \neg c, \neg a\}.$$

Some of the semantics considered in this chapter infer b from $\mathcal{P}$. Is that reasonable? (There is no consensus answer to this question among logic programmers.)

This sort of program has been suggested for expressing diagnosis questions. (See, e.g., (Kakas and Mancarella 1990).) For example, in program $\mathcal{P}$, let a stand for "My dog does not have fleas"; b for "My dog has fleas"; and p for "My dog scratches frequently." The first two rules seem to capture only the idea "Either my dog has fleas or he doesn't." In some of the semantics discussed, the last rule captures the assertion "My dog scratches frequently" but makes the semantics search for a reason why he scratches frequently, causing the semantics to infer b. Is this a reasonable use of the rules, or is the ability to use the rules to make this diagnosis an example of a "bug becoming a feature"?

Exercise 1. For each logic program presented in Examples 5.1 and 5.2, and for each logic programming semantics discussed later in this chapter, identify the set of ground literals inferred by the logic programming semantics from the program. (Warning: some of the semantics are not defined on some of the programs.)

6 A semantics survey

In this section we describe several semantics that have been proposed for normal logic programs. We shall use the following notation: for any semantics $\mathcal{S}$ and any logic program $\mathcal{P}$, $\textit{Inferences}_\mathcal{S}(\mathcal{P})$ is the set of inferences made by semantics $\mathcal{S}$ from $\mathcal{P}$. When the semantics $\mathcal{S}$ is obvious from context, we shall write simply $\textit{Inferences}(\mathcal{P})$.

6.1 Classical logic

Obviously, classical logic was not created explicitly for logic programming. But it is also obviously appropriate to consider it. Here the $\leftarrow$ is interpreted classically. Infer all the consequences of the set $\mathcal{P}$ of sentences in classical logic.

Proposition 6.1 *For $\mathcal{P}$ any normal logic program: $\mathcal{P}$ is satisfied in the model $\mathbf{B}_\mathcal{P}$ itself (i.e., the model where every atomic sentence is true). Hence in classical logic we cannot infer any negative atoms from any normal logic program.*

Thus classical logic does not exhibit negation as failure for logic programs.

6.2 Minimal model semantics

Recall that an element x of a partial ordering $(X, \preceq)$ is *minimal* if

$$\forall y \in X (y \preceq x \rightarrow y = x),$$

and x is the *minimum* element of $(X, \preceq)$ if

$$\forall y \in X (x \preceq y).$$

Of course, if a partial ordering has a minimum element x, x is also the unique minimal element.

In the context of logic programming, people consider the set of all Herbrand models of a program $\mathcal{P}$. Recall that Herbrand models are identified with subsets of the Herbrand base of $\mathcal{P}$. Thus Herbrand models are naturally partially ordered by $\subseteq$. There is a clear correlation with negation as failure: if $\mathcal{M} \subset \mathcal{N}$, then every atomic sentence which is false in $\mathcal{N}$ is also false in $\mathcal{M}$ – and at one least additional atomic sentence is false too. Thus

$\mathcal{M}$ seems to come closer to exhibiting negation as failure than $\mathcal{N}$ does.

By Theorem 2.7, each definite program has a minimum Herbrand model. For more general sets of sentences, including normal logic programs, minimum models usually do not exist. But the set of all minimal Herbrand models is still felt to be significant.

Definition 6.2 *The minimal model semantics treats the minimal Herbrand models of any set S of sentences as the intended models of S. The semantics infers from S, depending upon the version of the definition, either*

- *the set of all sentences true in all minimal Herbrand models of S, or*
- *the set of all ground literals (both positive and negative) true in all minimal Herbrand models.*

Example 6.3

(1) Let $\mathcal{P} = \{p \leftarrow p\}$. Then $\mathcal{P}$ has two models, $\{p\}$ and $\emptyset$. Program $\mathcal{P}$ is a definite program, and $\emptyset$ is the minimum model, and hence the unique minimal model. So the minimal model semantics infers $\neg p$.

(2) Let $\mathcal{P} = \{a \leftarrow \neg b\}$. Then $\mathcal{P}$ has three models, $\{a\}$, $\{b\}$, and $\{a, b\}$. The first two models are minimal; the third is not. Hence the semantics infers neither a nor $\neg a$, neither b nor $\neg b$, though it infers $a \vee b$ in the version where it infers all *sentences* true in all minimal models.

(3) Consider the slightly more complicated program

$$\mathcal{P} = \{a \leftarrow \neg b, \; z \leftarrow \neg a\}.$$

There are $2^3 = 8$ interpretations:

$$\emptyset, \; \{a\}, \; \{b\}, \; \{z\}, \; \{a,b\}, \; \{a,z\}, \; \{b,z\}, \; \{a,b,z\}.$$

Of these, $\{a\}$, $\{a,b\}$, $\{a,z\}$, $\{b,z\}$, and $\{a,b,z\}$ are models of $\mathcal{P}$; $\{a\}$ and $\{b,z\}$ are minimal. Thus no literals are inferred by the semantics.

(4) Consider the example

$$\mathcal{P} = \{a \leftarrow \neg b, \; b \leftarrow \neg a, \; c \leftarrow a, \; c \leftarrow b\}.$$

The models are $\{a,c\}$, $\{b,c\}$, $\{a,b,c\}$. The first two are minimal. So the minimal model semantics infers c, but it infers neither a nor $\neg a$ and neither b nor $\neg b$.

(5) Let $\mathcal{P} = \{a \leftarrow \neg a, \; b \leftarrow\}$. Then $\mathcal{P}$ has just one model, $\{a,b\}$, so the minimal model semantics infers a and b.

There are sets S of sentences which have Herbrand models but no minimal Herbrand models. The minimal model semantics would treat such a set S as contradictory, even though S does have Herbrand models. Some researchers find that to be an undesirable feature of the semantics. But in fairly broad circumstances, this difficulty cannot arise:

Theorem 6.4

(1) *Suppose S is a finite set of sentences with no function symbols (thus with a finite Herbrand universe). If $\mathcal{N}$ is an Herbrand model of S, then there is a minimal Herbrand model $\mathcal{M}$ of S with $\mathcal{M} \subseteq \mathcal{N}$.*

(2) *Suppose S is a set of universal sentences (e.g., a normal logic program). If $\mathcal{N}$ is an Herbrand model of S, then there is a minimal Herbrand model $\mathcal{M}$ of S with $\mathcal{M} \subseteq \mathcal{N}$.*

Exercise 2. Prove Theorem 6.4.

6.3 Perfect models of stratified programs

The van Emden Kowalski semantics is defined only for definite logic programs. A somewhat larger class of normal programs that has been important is the class of *stratified programs* (which we define below). On these there is a well-accepted semantics, called the perfect model semantics. Stratified programs and this semantics were introduced by (Apt et al. 1988) and (Van Gelder 1989a), with earlier related work in (Chandra and Harel 1985). Later researchers (see (Przymusiński 1988)) have applied the same semantics to larger classes of programs. In any case, many interesting logic programs are stratified; in fact, for a time it was suspected by many that all interesting logic programs are stratified.

Like the van Emden Kowalski semantics, the perfect model semantics for stratified programs picks out a *unique* Herbrand model, called the perfect model, as the intended model for the logic program. Like the minimum model in the van Emden Kowalski semantics, the perfect model is widely accepted as capturing exactly the intuitions of negation as failure – in this case for stratified programs. With the minimal model semantics and the remaining four semantics we shall discuss, on the other hand, (1) no single intended model is picked out, (2) the semantics are defined for all normal logic programs (though sometimes some of them give contradictory inferences) and (3) there is lack of agreement about whether the semantics correctly capture negation as failure intuitions.

Definition 6.5 *A relation symbol or proposition letter is defined in a rule or a program if it appears in the head of the rule or of some rule in the program; it is used positively in a rule or a program if it appears in a positive subgoal of the rule or of some rule in the program; it is used negatively in a rule or a program if it appears in a negative subgoal of the rule or of some rule in the program.*

Definition 6.6 *A normal program $\mathcal{P}$ is stratified if it is the union*

$$\mathcal{P}_1 \cup \mathcal{P}_2 \cup \cdots \cup \mathcal{P}_n$$

of programs, where (1) each relation and proposition letter is defined in at most one of the $\mathcal{P}_i$'s, (2) for $1 \leq i \leq n$, each relation and proposition letter defined in $\mathcal{P}_i$ is used positively only in $\mathcal{P}_j$'s for $j \geq i$ and used negatively only in $\mathcal{P}_j$'s for $j > i$, and (3) no $\mathcal{P}_i$ is empty. (This last requirement is just a convenience. We shall later, for notational convenience in defining the semantics, add a $\mathcal{P}_0$ where $\mathcal{P}_0$ is always empty.) The sequence $(\mathcal{P}_1, \mathcal{P}_2, \ldots, \mathcal{P}_n)$ is a stratification of $\mathcal{P}$, and $\mathcal{P}_1, \mathcal{P}_2, \ldots \mathcal{P}_n$ are strata.

Note that, in particular, if $(\mathcal{P}_1, \mathcal{P}_2, \ldots, \mathcal{P}_n)$ is a stratification of $\mathcal{P}$, then $\mathcal{P}_1$ is a definite program.

Example 6.7 We elaborate Example 5.1(2), describing a business where no supervisor may hire his (her) own relatives or inlaws. Let $\mathcal{P} = \mathcal{P}_1 \cup \mathcal{P}_2 \cup \mathcal{P}_3$, where

$\mathcal{P}_1 =$
$\{married(\text{Henry}, \text{Catherine}) \leftarrow$
$\phantom{\{}married(\text{Catherine}, \text{Henry}) \leftarrow$
$\phantom{\{}father(\text{Henry}, \text{Mary}) \leftarrow$
$\phantom{\{}mother(\text{Catherine}, \text{Mary}) \leftarrow$
$\phantom{\{}married(\text{Henry}, \text{Anne}) \leftarrow$
$\phantom{\{}married(\text{Anne}, \text{Henry}) \leftarrow$
$\phantom{\{}father(\text{Henry}, \text{Elizabeth}) \leftarrow$
$\phantom{\{}mother(\text{Anne}, \text{Elizabeth}) \leftarrow$
$\phantom{\{}married(\text{Isabella}, \text{Ferdinand}) \leftarrow$
$\phantom{\{}married(\text{Ferdinand}, \text{Isabella}) \leftarrow$
$\phantom{\{}mother(\text{Isabella}, \text{Catherine}) \leftarrow$
$\phantom{\{}father(\text{Ferdinand}, \text{Catherine}) \leftarrow$
$\phantom{\{}ancestor(X, X) \leftarrow$
$\phantom{\{}ancestor(X, Z) \leftarrow ancestor(X, Y), mother(Y, Z)$
$\phantom{\{}ancestor(X, Z) \leftarrow ancestor(X, Y), father(Y, Z)$
$\phantom{\{}relative(Y, Z) \leftarrow ancestor(X, Y), ancestor(X, Z)\}$

$\mathcal{P}_2 =$
$\{inlaw(X, Z) \leftarrow relative(X, Y), married(Y, Z), \neg relative(X, Z)$
$\phantom{\{}inlaw(X, Y) \leftarrow inlaw(Y, X)\}$

$\mathcal{P}_3 =$
$\{mayHire(X, Y) \leftarrow \neg relative(X, Y), \neg inlaw(X, Y)\}$.

Note that $\mathcal{P}_1$ is a definite program, defining a set

$$\mathcal{L}_1 = \{married, father, mother, ancestor, relative\}$$

of relations; hence, the van Emden Kowalski semantics for $\mathcal{P}_1$ prescribes interpretations of the symbols in $\mathcal{L}_1$.

Program $\mathcal{P}_2$ now uses the symbols defined in $\mathcal{P}_1$, but it does not add to their definitions. At this point, the intuition of the perfect model semantics is that $\mathcal{P}_1$ defined its symbols, and that $\mathcal{P}_2$ will use those definitions to define its additional symbols. For example, Ferdinand and Isabella are inlaws of Henry but not of Mary (because she is their relative), Anne, or

Elizabeth. Then $\mathcal{P}_3$ will use the definitions of symbols defined in $\mathcal{P}_1$ and $\mathcal{P}_2$ to define still more symbols. For example, Ferdinand may hire Anne and Elizabeth, but not Mary (a relative) or Isabella or Henry (both inlaws).

To define the perfect model for a stratified program, we need only make this intuition precise.

Given a ground normal logic program $\mathcal{Q}$ and an interpretation $\mathcal{M}$ for a subset $\mathcal{L}$ of the relation and proposition symbols in $\mathcal{Q}$, we replace atomic formulas of $\mathcal{L}$ with their truth values in $\mathcal{M}$ and then simplify the result. We formally define this as follows. For any set $\mathcal{L}$ of relation symbols and proposition letters, a ground atom $r(t_1, \ldots, t_m)$ is an *atom of $\mathcal{L}$* if $r \in \mathcal{L}$, and a proposition letter a is an *atom of $\mathcal{L}$* if $a \in \mathcal{L}$.

Definition 6.8 *Let $\mathcal{Q}$ be a ground normal logic program; $\mathcal{L}$, a set of relation symbols and proposition letters; and $\mathcal{M}$, a set of ground atoms of $\mathcal{L}$. The simplification of $\mathcal{Q}$ over $(\mathcal{L}, \mathcal{M})$ is the normal logic program formed from $\mathcal{Q}$ as follows:*

(1) *Whenever a ground atom of $\mathcal{L}$ appears as a subgoal in a rule of $\mathcal{Q}$, replace it with its truth value in $\mathcal{M}$.*

(2) *In the result of step (1), simplify the formulas: delete each subgoal of true or $\neg$false (since such a subgoal is trivially satisfied) and delete each rule with a subgoal of false or $\neg$true (since such a rule is vacuously satisfied).*

Note that if no atom of $\mathcal{L}$ appears in the head of any rule in $\mathcal{Q}$, then the simplification of $\mathcal{Q}$ over $(\mathcal{L}, \mathcal{M})$ is a normal logic program involving only atomic sentences not in $\mathcal{L}$.

Now we continue with the informal description of the perfect model for the ancestry example above. First, construct the ground instantiations $\mathcal{P}_1^g$, $\mathcal{P}_2^g$, and $\mathcal{P}_3^g$ of $\mathcal{P}_1$, $\mathcal{P}_2$, and $\mathcal{P}_3$, over the Herbrand universe $\mathbf{U}_\mathcal{P}$. Second, construct the minimum model $\mathcal{M}_1$ of $\mathcal{P}_1^g$. Then the simplification of $\mathcal{P}_2^g$ over $(\mathcal{L}_1, \mathcal{M}_1)$ is a definite program, consisting of rules such as

$$inlaw(Isabella, Henry) \leftarrow$$

(with all subgoals deleted in the process of simplification) and

$$inlaw(Isabella, Anne) \leftarrow inlaw(Anne, Isabella).$$

Form its minimum model $\mathcal{M}_2$, and let $\mathcal{L}_2 = \{inlaw\}$. Finally, the simplification of $\mathcal{P}_3^g$ over $(\mathcal{L}_1 \cup \mathcal{L}_2, \mathcal{M}_1 \cup \mathcal{M}_2)$ is a definite program, so form its minimum model $\mathcal{M}_3$. The perfect model of $\mathcal{P}$ is $\mathcal{M}_1 \cup \mathcal{M}_2 \cup \mathcal{M}_3$.

We say that two Herbrand interpretations $\mathcal{M}_1, \mathcal{M}_2$ *agree on symbols in $\mathcal{L}$* if, for each ground atom α of $\mathcal{L}$, $\alpha \in \mathcal{M}_1$ if and only if $\alpha \in \mathcal{M}_2$.

Definition 6.9 *Suppose $(\mathcal{P}_1, \mathcal{P}_2, \ldots, \mathcal{P}_n)$ is a stratification of a normal logic program $\mathcal{P}$, and suppose that $\mathcal{L}$ is the set of relation and function symbols appearing in $\mathcal{P}$. Then the corresponding partition of $\mathcal{L}$ is the se-*

quence $(\mathcal{L}_0, \mathcal{L}_1, \mathcal{L}_2, \ldots, \mathcal{L}_n)$ where each $\mathcal{L}_j$, for $j > 0$, is the set of relation symbols and proposition letters in $\mathcal{P}$ which are defined in $\mathcal{P}_j$, and $\mathcal{L}_0$ is the set of relation symbols and proposition letters of $\mathcal{P}$ which are not defined in $\mathcal{P}$. We shall also say that $(\mathcal{P}_0, \mathcal{P}_1, \mathcal{P}_2, \ldots, \mathcal{P}_n)$ is the stratification, where $\mathcal{P}_0 = \emptyset$ (we shall always imply $\mathcal{P}_0 = \emptyset$, the part of $\mathcal{P}$ defining symbols in $\mathcal{L}_0$); this will allow more uniform notation.

Note that, for a stratification $(\mathcal{P}_0, \mathcal{P}_1, \mathcal{P}_2, \ldots, \mathcal{P}_n)$ of a normal logic program $\mathcal{P}$ and for the corresponding partition $(\mathcal{L}_0, \mathcal{L}_1, \mathcal{L}_2, \ldots, \mathcal{L}_n)$, for each i, $0 \leq i \leq n$, no symbol r or a in $\mathcal{L}_i$ appears in $\mathcal{P}_0, \mathcal{P}_1, \ldots, \mathcal{P}_{i-1}$.

The perfect model of $\mathcal{P}$ is an Herbrand model, so it is built upon the Herbrand preinterpretation $\mathbf{U}_\mathcal{P}$. For the remainder of this subsection, by an Herbrand model of any $\mathcal{P}_i$ or its simplification, we mean one expanding $\mathbf{U}_\mathcal{P}$, the Herbrand universe for the entire program, not just $\mathbf{U}_{\mathcal{P}_i}$. Note that, just as with the van Emden Kowalski semantics, if any $\mathcal{P}_i$ or its simplification is a definite clause program, it has a minimum model (constructed by the immediate analogue of the $\mathbf{T}_\mathcal{P}$ construction given earlier, generalizing that construction to involve a potentially larger set of function and constant symbols).

Lemma 6.10 *Suppose that $(\mathcal{P}_1, \mathcal{P}_2, \ldots, \mathcal{P}_n)$ is a stratification of a normal logic program $\mathcal{P}$, let $(\mathcal{L}_0, \mathcal{L}_1, \mathcal{L}_2, \ldots, \mathcal{L}_n)$ be the corresponding partition of the set of relation symbols and proposition letters appearing in $\mathcal{P}$, and let $\mathcal{P}_0 = \emptyset$. For any i, $0 \leq i \leq n$, let $\mathcal{L}_{<i} = \mathcal{L}_0 \cup \cdots \cup \mathcal{L}_{i-1}$, and let $\mathcal{P}_{<i} = \mathcal{P}_0 \cup \cdots \cup \mathcal{P}_{i-1}$, Let $\mathcal{M}_{<i}$ be an Herbrand model of $\mathcal{P}_{<i}$. For $i \leq j \leq n$, let $\mathcal{Q}_j^i$ be the simplification of the ground instantiation of $\mathcal{P}_j$ over $(\mathcal{L}_{<i}, \mathcal{M}_{<i})$, and let $\mathcal{Q}^i$ be the simplification of the ground instantiation of $\mathcal{P} - \mathcal{P}_{<i}$ over $(\mathcal{L}_{<i}, \mathcal{M}_{<i})$. Then*

(1) *$\mathcal{Q}_i^i$ is a (possibly empty) definite program.*
(2) *$(\mathcal{Q}_i^i, \ldots, \mathcal{Q}_n^i)$ is a stratification of $\mathcal{Q}^i$.*
(3) *$\{\mathcal{M} \cup \mathcal{M}_{<i} : \mathcal{M}$ is an Herbrand model of $\mathcal{Q}^i\}$ is exactly the set of Herbrand models $\mathcal{N}$ of $\mathcal{P}$ which agree with $\mathcal{M}_{<i}$ on symbols in $\mathcal{L}_1$.*

Proof Left to reader.

The formal definition of the term *perfect model* is by induction on the number of strata in the program. Lemma 6.10 shows that the process is well-defined. The initial definition is given in terms of not only the program but also the stratification.

Definition 6.11 *Continue all notation from the lemma above. Let $\mathcal{M}_{<0}$, $\mathcal{L}_{<0} = \emptyset$.*

- *For $0 \leq i \leq n$, let $\mathcal{M}_i$ be the minimum Herbrand model for $\mathcal{Q}_i^i$. (Recall that $\mathcal{Q}_i^i$ is the simplification of the ground instantiation of $\mathcal{P}_i$ over $(\mathcal{M}_{<i}, \mathcal{L}_{<i})$.)*
- *For $1 \leq i \leq n+1$, let $\mathcal{M}_{<i} = \mathcal{M}_0 \cup \mathcal{M}_1 \cup \cdots \cup \mathcal{M}_{i-1}$.*

Then $\mathcal{M}_{<n+1}$ is the perfect model of $\mathcal{P}$ under stratification $(\mathcal{P}_1, \mathcal{P}_2, \ldots, \mathcal{P}_n)$.

Observe that, in the definition above, for each $i > 0$, $\mathcal{M}_{<i+1}$ is the perfect model of $\mathcal{P}_1 \cup \cdots \cup \mathcal{P}_i$ under the stratification $(\mathcal{P}_1, \mathcal{P}_2, \ldots, \mathcal{P}_i)$.

Example 6.12 Consider again Example 6.3.

(1) Let $\mathcal{P} = \{p \leftarrow p\}$. $\mathcal{P}$ consists of one rule and is definite; hence it is stratified with one stratum. Its perfect model is $\emptyset$, which was also its unique minimal model.

(2) Let $\mathcal{P} = \{a \leftarrow \neg b\}$. The program is stratified with a single stratum. Hence we have $\mathcal{P}_0 = \emptyset$, $\mathcal{L}_0 = \{b\}$, $\mathcal{P}_1 = \mathcal{P}$, and $\mathcal{L}_1 = \{a\}$.
$\mathcal{M}_0$ is the minimum model of $\mathcal{P}_0$, so $\mathcal{M}_0 = \emptyset$. Now simplify $\mathcal{P}_1$ over $(\mathcal{L}_0, \mathcal{M}_0)$: since $b \in \mathcal{L}_0$ but $b \notin \mathcal{M}_0$, b evaluates to *false*. So, substituting, we get the rule $a \leftarrow \neg false$, which simplifies to $a \leftarrow$. Thus $\mathcal{M}_1 = \{a\}$. Hence the perfect model is $\{a\}$; b is inferred to be false and a is inferred to be true.
Note that the perfect model is one of the two minimal models of $\mathcal{P}$.

(3) Let $\mathcal{P} = \{a \leftarrow \neg b, \ z \leftarrow \neg a\}$. There are two strata, $\mathcal{P}_1 = \{a \leftarrow \neg b\}$ and $\mathcal{P}_2 = \{z \leftarrow \neg a\}$. As above, the perfect model for $\mathcal{P}_1$ is $\{a\}$, with $\mathcal{L}_1 = \{a, b\}$. Substituting truth values into $\mathcal{P}_2$, we get the one rule $z \leftarrow \neg true$, which is vacuous and eliminated during simplification. Hence $\mathcal{M}_2 = \emptyset$, and the perfect model is $\mathcal{M}_0 \cup \mathcal{M}_1 \cup \mathcal{M}_2 = \{a\}$.
Again, the perfect model is one of the two minimal models.

(4) The program
$$\mathcal{P} = \{a \leftarrow \neg b, \ b \leftarrow \neg a, \ c \leftarrow a, \ c \leftarrow b\}$$
is not stratified, since, if it were stratified, each of the two rules $a \leftarrow \neg b$ and $b \leftarrow \neg a$ would have to be in a lower-numbered stratum than the other. Thus the semantics is not defined for this program.

(5) The program $\mathcal{P} = \{a \leftarrow \neg a, \ b \leftarrow\}$ is not stratified, since the rule that defines a also uses a negatively. The stratified semantics is not defined for this program either.

Lemma 6.13 *Suppose a logic program $\mathcal{P}$ has two stratifications,*
$$(\mathcal{P}_1, \mathcal{P}_2, \ldots, \mathcal{P}_m) \quad \text{and} \quad (\mathcal{Q}_1, \mathcal{Q}_2, \ldots, \mathcal{Q}_m).$$
Then the perfect models of $\mathcal{P}$ under the two stratifications are the same.

Exercise 3. Prove Lemma 6.13.

Definition 6.14 *Let $\mathcal{P}$ be a stratified logic program. Its perfect model is its perfect model under any stratification. The stratified semantics picks the perfect model as the intended model of any stratified logic program. Thus it infers all sentences (or all ground literals) true in the perfect model.*

Exercise 4. Prove that for any stratified logic program $\mathcal{P}$, the perfect model of $\mathcal{P}$ is a minimal model of $\mathcal{P}$. Thus, for stratified programs, the perfect model semantics makes all inferences made by the minimal model semantics.

6.4 2-valued program completion semantics

In the remainder of this section, we define four semantics which apply to all logic programs $\mathcal{P}$, though sometimes they may make contradictory inferences. The first is the 2-valued program completion semantics of Clark (1978), or rather, a commonly used minor modification of Clark's semantics.

Recall that in discussing stratified programs we spoke of a rule "defining" the relation symbol or proposition letter in its head. Clark treated the set of all rules defining a symbol as *the definition of* the symbol, thus justifying a translation: if a program contains two rules defining a proposition letter a, say $a \leftarrow \phi$ and $a \leftarrow \psi$, Clark's translation of the rules defining a is the sentence

$$a \leftrightarrow \phi \vee \psi$$

asserting that a is true if *and only if* the body of some rule with a as its head is true. Thus if proposition letter b is the head of no rules, Clark's translation is $b \leftrightarrow \text{\textit{false}}$, since *false* is the disjunction of 0 formulas.

In the case of first order logic, Clark's translation is a bit more complicated; we just give an example.[1] Suppose binary relation r appears in the head of just three rules in a program:

$$r(1,1) \leftarrow, \quad r(X+1, Y) \leftarrow r(X, Y+1), \quad r(X,Y) \leftarrow r(X+Y, 0).$$

Clark's translation of these rules is then

$$\begin{aligned} r(X,Y) \quad &\leftrightarrow \quad (X = 1 \wedge Y = 1) \\ &\vee \quad \exists X', Y'(X = X' + 1 \wedge Y' = Y + 1 \wedge r(X', Y')) \\ &\vee \quad r(X+Y, 0). \end{aligned}$$

In this paper we formulate the semantics instead using the the ground instantiation of a logic program, where the ground instantiation may be treated as a possibly infinite propositional logic program. The translation is performed for each ground atom. If a ground atom a is the head of finitely many rules, $a \leftarrow \phi_i$ for $i = 1, \ldots, n$, the translation is

$$a \leftrightarrow \phi_1 \vee \phi_2 \vee \cdots \vee \phi_n.$$

If ground atom a is the head if infinitely many rules, say rules $a \leftarrow \phi_n$ for

---

[1] For intuitive clarity we use + as an infix operator, as in ordinary mathematics, not as a prefix (binary) function – or even a prefix (ternary) relation – as would be correct for logic programming.

every positive integer n, then the translation is

$$a \leftrightarrow \bigvee_{i=1}^{\infty} \phi_i.$$

This formula is, of course, of infinite length, so this approach takes us outside ordinary logic.

The set of all such translations – one for each ground atom of the program – is called the *completion* of $\mathcal{P}$. The models of the completion are taken to be the intended models of $\mathcal{P}$.

Definition 6.15 *From a program, the 2-valued program completion semantics infers all sentences (or ground literals) true in all Herbrand models of the completion of the program.*[2]

One can avoid mentioning infinitary logic by considering fixed points of operators defined in first order logic.

Definition 6.16 *Suppose $\mathcal{P}$ is a normal logic program and let $\mathcal{M}$ be an Herbrand interpretation for $\mathcal{P}$, that is, a set of ground atoms of $\mathcal{P}$. Let*

$\mathbf{T}_\mathcal{P}(\mathcal{M}) = \{\alpha' : \text{for some ground instance } \alpha' \leftarrow \beta'_1, \ldots, \beta'_n \text{ of a rule of } \mathcal{P}, \text{ each } \beta'_i \text{ is true in } \mathcal{M}\}.$

Suppose $\mathcal{P}$ is a normal logic program and $\mathcal{M}$ is an Herbrand interpretation for $\mathcal{P}$. Observe that $\mathcal{M}$ is a model of $\mathcal{P}$ if and only if $\mathbf{T}_\mathcal{P}(\mathcal{M}) \subseteq \mathcal{M}$. Recall that a *fixed point* of a function f is an object x where $f(x) = x$.

Proposition 6.17 *An Herbrand interpretation $\mathcal{M}$ for a normal logic program $\mathcal{P}$ is a model of the completion of $\mathcal{P}$ if and only if it is a fixed point of $\mathbf{T}_\mathcal{P}$. The 2-valued program completion semantics infers from $\mathcal{P}$ all sentences (or ground literals) true in all fixed points of $\mathbf{T}_\mathcal{P}$.*

Proof First suppose $\mathcal{M}$ is a fixed point of $\mathbf{T}_\mathcal{P}$. Suppose a ground atom a is the head of ground rules $a \leftarrow \phi_i$. Since $\mathcal{M}$ is a fixed point, $a \in \mathcal{M}$ if and only if some ϕ_i is true in $\mathcal{M}$. But then $a \in \mathcal{M}$ if and only if $\bigvee_i \phi_i$ is true in $\mathcal{M}$. Thus

$$a \leftrightarrow \bigvee_i \phi_i$$

is true in $\mathcal{M}$. The converse is proved similarly. The final statement is an immediate corollary. ∎

As we build upon the idea of the completion in later sections, we concentrate upon this fixed point characterization of the 2-valued program completion semantics.

---

[2] In the case of logic programs with function symbols, Clark's original formulation may make somewhat fewer inferences.

Example 6.18 Consider again Example 6.3.

(1) Let $\mathcal{P} = \{p \leftarrow p\}$. Its completion is $\{p \leftrightarrow p\}$, which is a tautology. Hence there are two models of the completion, $\emptyset$ and $\{p\}$. (That $\emptyset$ and $\{p\}$ are both fixed points of $\mathcal{P}$ is also easily established directly from the definition of $\mathbf{T}_\mathcal{P}$.) So neither p nor $\neg p$ is inferred.

This contrasts with the van Emden Kowalski semantics, the minimal model semantics, and the perfect model semantics, all of which infer $\neg p$.

(2) Let $\mathcal{P} = \{a \leftarrow \neg b\}$. Its completion is $\{a \leftrightarrow \neg b,\ b \leftrightarrow \text{false}\}$, which has one model, $\{a\}$. (Similarly, one can show that, since no rule has head b, b is not in any fixed point of $\mathbf{T}_\mathcal{P}$, and since b is not in any fixed point, a is in every fixed point; and finally, one can easily check that $\{a\}$ is a fixed point.)

So the semantics infers a to be true and b to be false, just as did the perfect model semantics. Recall that the minimal model semantics inferred no literals.

(3) Let $\mathcal{P} = \{a \leftarrow \neg b,\ z \leftarrow \neg a\}$. Since there is no rule with head b, b cannot be in any fixed point of $\mathbf{T}_\mathcal{P}$. Since b cannot be in any fixed point, a must be in every fixed point. Since a must be in every fixed point, and the only rule with z in its head is $z \leftarrow \neg a$, z cannot be in any fixed point. Finally, one can easily check that $\{a\}$ is a fixed point.

Hence the semantics infers a, $\neg b$, and $\neg z$, just as did the perfect model semantics. The minimal model semantics inferred no literals.

(4) $\mathcal{P} = \{a \leftarrow \neg b,\ b \leftarrow \neg a,\ c \leftarrow a,\ c \leftarrow b\}$. Since every fixed point of $\mathbf{T}_\mathcal{P}$ is a model of $\mathcal{P}$, the fixed points must be among $\{a, c\}$, $\{b, c\}$, and $\{a, b, c\}$. One can easily check that the first two are fixed points and the third is not. (Also, note that the completion is equivalent to $\{a \underline{\vee} b,\ c \leftrightarrow a \vee b\}$.) So the 2-valued program completion semantics infers c but neither a nor $\neg a$ and neither b nor $\neg b$.

(5) For $\mathcal{P} = \{a \leftarrow \neg a,\ b \leftarrow\}$, the completion of $\mathcal{P}$ is $\{a \leftrightarrow \neg a,\ b \leftrightarrow \text{true}\}$, which is inconsistent.

(We could also check that there are four interpretations, $\emptyset$, $\{a\}$, $\{b\}$, and $\{a, b\}$. Inspection shows that for any interpretation $\mathcal{M}$, $a \in \mathcal{M}$ if and only if $a \notin \mathbf{T}_\mathcal{P}(\mathcal{M})$, so there are no fixed points of $\mathbf{T}_\mathcal{P}$.) Thus a, $\neg a$, b, and $\neg b$ are all (vacuously) true in all models of the completion of $\mathcal{P}$, and the 2-valued program completion semantics infers all four literals.

Here the 2-valued program completion semantics treats the rule $a \leftarrow \neg a$ as defining a, and that definition is nonsensical.

While working with the examples of Section 5, compare the 2-valued program completion semantics to the perfect model semantics (on those programs which are stratified) and to the minimal model semantics. Look in particular for examples where, as with Example 6.18(1), the semantics makes weak inferences and for examples where no fixed points exist, and thus where the semantics makes inconsistent inferences. Note that, in our definition of the program completion semantics, we did not assume that the program $\mathcal{P}$ is finite. This will be important later.

6.5   3-valued program completion semantics

Fitting (1985) proposed an interpretation of Clark's semantics in a 3-valued logic, with truth values *true*, *false*, and $\bot$. A natural interpretation of the three truth values is epistemic: a is assigned the truth value *true* or *false* if a is *believed to be* true or false; a is assigned the truth value $\bot$ if a is neither believed to be true nor believed to be false. In later papers Fitting also includes a fourth truth value, $\top$, representing that a proposition is believed to be both true and false. In spite of the fact that we shall represent all four truth values, we shall follow earlier tradition in calling the semantics the 3-valued program completion semantics.

Though we shall not use them explicitly, we also include the truth tables for the 3-valued logic, since they provide justification for some of our general comments.

$\neg$	
t	f
f	t
$\bot$	$\bot$

$\wedge$	t	f	$\bot$
t	t	f	$\bot$
f	f	f	f
$\bot$	$\bot$	f	$\bot$

$\vee$	t	f	$\bot$
t	t	t	t
f	t	f	$\bot$
$\bot$	t	$\bot$	$\bot$

A formula $\forall X \phi(X)$ has truth value *true* if, for every X in the universe, ϕ is true of X; it has truth value *false* if, for some X in the universe, ϕ is false of X; otherwise has truth value $\bot$. Satisfaction for $\exists$ is defined analogously.

In our formal treatment we shall not use the 3-valued logic. We shall represent the same epistemic notions by creating an extra relation symbol $r_?$ for each relation symbol r and an extra proposition letter $a_?$ for each proposition letter a, much in the style of (Fitting 1993) and (Schlipf 1991). If a denotes an atom $r(t_1, \ldots, t_m)$, we shall refer to the atom $r_?(t_1, \ldots, t_m)$ as $a_?$. The *understood* meanings are given below:

symbol	understood meaning
a	original atom a is believed to be true
$a_?$	original atom a is not believed to be false.

Suppose a program contains four proposition letters a, b, c, and d. An interpretation $\mathcal{M} = \{a, a_?, b_?, d\}$ represents that a is true (both believed to be true and not believed to be false), that b is neither believed to be true nor believed to be false, corresponding to truth value $\bot$, that c is false, and that d is believed to be both true and false, corresponding to truth value

Definition 6.19 *For $\mathcal{P}$ a normal logic program, let $\mathcal{P}_3$ be the logic program containing the following rules: for each rule*

$$a \leftarrow b_1, \ldots, b_k, \neg c_1, \ldots, \neg c_m$$

of $\mathcal{P}$, where a, the b_i's, and the c_j's are all positive literals, $\mathcal{P}_3$ has two rules:

$$a \leftarrow b_1, \ldots, b_k, \neg c_{1?}, \ldots, \neg c_{m?}$$
$$a_? \leftarrow b_{1?}, \ldots, b_{k?}, \neg c_1, \ldots, \neg c_m$$

The 3-valued program completion semantics infers all ground literals of $\mathcal{P}$ true in all fixed points of $\mathbf{T}_{\mathcal{P}_3}$.[3]

Note the fairly obvious intuitions of the two formulas in the translation, based upon the understood meanings above. The first is based upon what is believed: a is believed to be true if each b_i is believed to be true and no c_j is not believed to be false – i.e., every c_j is believed to be false. The second is based upon what is not explicitly believed: a is *possible* (i.e., $a_?$ is not false) if every b_i is possible and no c_i is believed to be true – i.e., every c_i is *possibly false*.

Example 6.20 Consider again Example 6.3.

(1) For $\mathcal{P} = \{p \leftarrow p\}$, $\mathcal{P}_3 = \{p \leftarrow p, \; p_? \leftarrow p_?\}$. All four interpretations are fixed points: $\{p, p_?\}$, $\{p_?\}$, $\{p\}$, and $\emptyset$. Thus the 3-valued program completion semantics infers neither p nor $\neg p$, just like the 2-valued program completion semantics. The two program completion semantics treat positive recursion similarly.

(2) For $\mathcal{P} = \{a \leftarrow \neg b\}$, $\mathcal{P}_3 = \{a \leftarrow \neg b_?, \; a_? \leftarrow \neg b\}$. Just as with the 2-valued program completion semantics, notice that $b, b_?$ can be in no fixed points, hence $a, a_?$ must be in every fixed point, and $\{a, a_?\}$ is a fixed point. So, like the 3-valued program completion semantics, the 2-valued program completion semantics infers a and $\neg b$.

(3) For $\mathcal{P} = \{a \leftarrow \neg b, \; z \leftarrow \neg a\}$,

$$\mathcal{P}_3 = \{a \leftarrow \neg b_?, \; a_? \leftarrow \neg b, \; z \leftarrow \neg a_?, \; z_? \leftarrow \neg a\}.$$

Just as with the 2-valued program completion semantics, notice that $b, b_?$ can be in no fixed point, so $a, a_?$ must be in every fixed point, so $z, z_?$ can be in no fixed point. And $\{a, a_?\}$ is a fixed point. So the 3-valued program completion semantics, like the 2-valued form, infers a, $\neg b$, and $\neg z$.

---

[3] We do not define here what it means for a sentence more complex than just a literal to be inferred in the 3-valued program completion semantics. This can however be defined straightforwardly using truth tables and definitions of the quantifiers given above and the understood meanings of the atoms a, $a_?$.

(4) For $\mathcal{P} = \{a \leftarrow \neg b, \ b \leftarrow \neg a, \ c \leftarrow a, \ c \leftarrow b\}$,

$$\mathcal{P}_3 = \{ \ a \leftarrow \neg b_?, \quad a_? \leftarrow \neg b, \quad b \leftarrow \neg a_?, \quad b_? \leftarrow \neg a,$$
$$c \leftarrow a, \quad c_? \leftarrow a_?, \quad c \leftarrow b, \quad c_? \leftarrow b_? \ \}$$

Observe that $\{a_?, b_?, c_?\}$ and $\{a, b, c\}$ are both fixed points of $\mathbf{T}_{\mathcal{P}_3}$. Thus the 3-valued program completion semantics infers no literals. The 2-valued program completion semantics could be thought of as inferring c by an argument by cases – either a or b must be true, and accordingly c must be true in either case. By allowing a third truth value $\bot$, the 3-valued version cannot make this inference.

(5) For $\mathcal{P} = \{a \leftarrow \neg a, \ b \leftarrow\}$,

$$\mathcal{P}_3 = \{a \leftarrow \neg a_?, \ a_? \leftarrow \neg a, \ b \leftarrow, \ b_? \leftarrow\}.$$

So, in any fixed point $\mathcal{M}$ of $\mathbf{T}_{\mathcal{P}_3}$, (1) $b \in \mathcal{M}$ and $b_? \in \mathcal{M}$, and (2) exactly one of $a, a_?$ is in $\mathcal{M}$. Conversely, it is apparent that $\{a, b, b_?\}$ and $\{a_?, b, b_?\}$ are both fixed points. So the 3-valued program completion semantics infers b and infers neither a nor $\neg a$.

Recall that, by contrast, the 2-valued program completion semantics made inconsistent inferences on this example. The 3-valued program completion semantics treats negative recursion quite differently than does the 2-valued version. In this case, one can interpret this as saying that the 2-valued program completion semantics treats the "definition" of a as nonsensical and passes off the entire program as nonsense, while the 3-valued version treats the "definition" of a as a difficulty it doesn't know how to resolve but notes that b follows in any case. (This example is from (Fitting 1985).)

Formalism for 4-valued semantics

Definition 6.21 *Suppose $\mathcal{P}$ is a normal logic program.*

(1) *Suppose $\mathcal{M}, \mathcal{N}$ are Herbrand interpretations for $\mathcal{P}_3$. Then $\mathcal{M} \sqsubseteq \mathcal{N}$ if for every ground atom a of $\mathcal{P}$,*
- *if $a \in \mathcal{M}$, then $a \in \mathcal{N}$, and*
- *if $a_? \notin \mathcal{M}$, then $a_? \notin \mathcal{N}$.*

(Note the natural intuition for $\sqsubseteq$: Suppose $\mathcal{M} \sqsubseteq \mathcal{N}$. Then for any ground atom a, if $\mathcal{M}$ asserts that a is believed to be true (i.e., $a \in \mathcal{M}$), so does $\mathcal{N}$. Furthermore, if $\mathcal{M}$ asserts that a is believed to be false (i.e., $a_? \notin \mathcal{M}$), so does $\mathcal{N}$. Thus $\sqsubseteq$ is an ordering of increasing beliefs).

(2) *An Herbrand interpretation $\mathcal{M}$ for $\mathcal{P}_3$ is consistent if, for each ground atom a of $\mathcal{P}$, if $a \in \mathcal{M}$ then $a_? \in \mathcal{M}$. The interpretation $\mathcal{M}$ is total if, for each ground atom a of $\mathcal{P}$, $a \in \mathcal{M}$ if and only if $a_? \in \mathcal{M}$.*

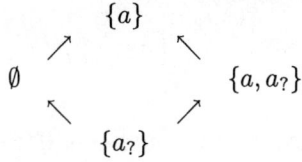

FIG. 1. Graph representing the $\sqsubseteq$ relation

(3) For an Herbrand interpretation $\mathcal{M}$ for $\mathcal{P}_3$, where a below varies over the ground atoms of $\mathcal{P}$, let

$$\widetilde{\mathcal{M}} = \{a : a_? \in \mathcal{M}\} \cup \{a_? : a \in \mathcal{M}\}.$$

Example 6.22 The Herbrand interpretations are partially ordered by $\sqsubseteq$. In figure 1 we picture the Herbrand interpretations for $\mathcal{P}_3$ where $\mathcal{P}$ contains only one proposition letter, a. In this figure, for interpretations $\mathcal{M}_0, \mathcal{M}_1$, $\mathcal{M}_0 \sqsubseteq \mathcal{M}_1$ if and only if there is a path in the graph, of length 0 or more, from $\mathcal{M}_0$ to $\mathcal{M}_1$. Here the interpretations $\emptyset$ and $\{a, a_?\}$ are total, and those two and also $\{a_?\}$ are consistent. Also, $\widetilde{\{a_?\}} = \{a\}$, $\widetilde{\emptyset} = \emptyset$, $\widetilde{\{a, a_?\}} = \{a, a_?\}$, and $\widetilde{\{a\}} = \{a_?\}$. The bottom interpretation, $\{a_?\}$ contains no information. The two middle interpretations, $\emptyset$ and $\{a, a_?\}$, contain total information, and the top interpretation, $\{a\}$, contains inconsistent information (representing that a is believed to be both true and false).

Note that if $\mathcal{P}$ contains (exactly) two proposition letters, a and b, then $\mathcal{P}_3$ contains $2^4 = 16$ elements. The bottom element of the partial ordering is $\{a_?, b_?\}$. Just above it are $\{a, a_?, b_?\}$, $\{a_?, b, b_?\}$, $\{b_?\}$, and $\{a_?\}$. All four are consistent and contain some information, but none are total.

Theorem 6.23 Let L be the set of all Herbrand interpretations of $\mathcal{P}_3$. Then:

(1) $(L, \sqsubseteq)$ is a complete lattice – call the join $\sqcup$ and the meet $\sqcap$.
(2) $\{a_? : a$ is a ground atom of $\mathcal{P}\}$ is the bottom element of $(L, \sqsubseteq)$ – call it $\bot_L$.
(3) $\{a : a$ is a ground atom of $\mathcal{P}\}$ is the top element of $(L, \sqsubseteq)$.
(4) $\mathbf{T}_{\mathcal{P}_3}$ is monotonic on $(L, \sqsubseteq)$, i.e., if $\mathcal{M}, \mathcal{N} \in L$ and $\mathcal{M} \sqsubseteq \mathcal{N}$, then $\mathbf{T}_{\mathcal{P}_3}(\mathcal{M}) \sqsubseteq \mathbf{T}_{\mathcal{P}_3}(\mathcal{N})$.
(5) $\mathbf{T}_{\mathcal{P}_3}$ has a $\sqsubseteq$-minimum fixpoint – call it $\mathbf{F}_\mathcal{P}$.
(6) The 3-valued program completion semantics never infers both a and $\neg a$ for any ground atom a of $\mathcal{P}$.
(7) $\mathbf{F}_\mathcal{P}$ can be constructed by transfinite induction:

$$\bot_L \sqsubseteq \mathbf{T}_{\mathcal{P}_3}(\bot_L) \sqsubseteq \mathbf{T}^2_{\mathcal{P}_3}(\bot_L) \sqsubseteq \cdots \sqsubseteq \mathbf{T}^\infty_{\mathcal{P}_3}(\bot_L) = \mathbf{F}_\mathcal{P},$$

where for a limit ordinal λ (such as the first infinite ordinal ω),

$\mathbf{T}_{\mathcal{P}_3}^\lambda(\perp_L)$ is defined to be

$$\bigsqcup_{\eta<\lambda} \mathbf{T}_{\mathcal{P}_3}^\alpha(\perp_L).$$

(8) For $\mathcal{M} \in L$, $\mathbf{T}_{\mathcal{P}_3}(\mathcal{M}) = \mathcal{M}$ if and only if $\mathbf{T}_{\mathcal{P}_3}(\widetilde{\mathcal{M}}) = \widetilde{\mathcal{M}}$.

(9) For a a ground atom of $\mathcal{P}$, $a \in \mathbf{F}_{\mathcal{P}}$ if and only if a is in every fixed point of $\mathbf{T}_{\mathcal{P}_3}$, and $a_? \in \mathbf{F}_{\mathcal{P}}$ if and only if a is in some fixed point of $\mathbf{T}_{\mathcal{P}_3}$.

(10) $\mathbf{F}_{\mathcal{P}}$ is consistent.

Consider again Examples 6.18(5) and 6.20(5). For that program the 2-valued program completion semantics makes contradictory inferences, while the 3-valued program completion semantics does not. This is generalized by part 6 above, which says that the 3-valued program completion semantics never makes inconsistent inferences.

Definition 6.24 *The transfinite inductive construction described in part 7 of the Theorem has standard notation:* $\mathbf{T}_{\mathcal{P}_3} \uparrow 0 = \perp_L$, *and, in general,* $\mathbf{T}_{\mathcal{P}_3} \uparrow \eta = \mathbf{T}_{\mathcal{P}_3}^\eta(\perp_L)$.[4]

Definition 6.25 *Let $\mathcal{P}$ be a normal logic program. For $\mathcal{M}$ an Herbrand interpretation for $\mathcal{P}$, let $\widehat{\mathcal{M}} = \mathcal{M} \cup \{a_? : a \in \mathcal{M}\}$.*

Note that an Herbrand interpretation $\mathcal{M}$ for $\mathcal{P}_3$ is total if and only if there is an interpretation $\mathcal{N}$ for $\mathcal{P}$ where $\mathcal{M} = \widehat{\mathcal{N}}$.

Theorem 6.26 *Let $\mathcal{P}$ be a normal logic program.*

(1) *For $\mathcal{M}$ any Herbrand interpretation for $\mathcal{P}$, $\mathbf{T}_{\mathcal{P}}(\mathcal{M}) = \mathcal{M}$ if and only if $\mathbf{T}_{\mathcal{P}_3}(\widehat{\mathcal{M}}) = \widehat{\mathcal{M}}$.*

(2) *Every ground literal of $\mathcal{P}$ true in all fixed points of $\mathbf{T}_{\mathcal{P}_3}$ is also true in all fixed points of $\mathbf{T}_{\mathcal{P}}$.*

(3) *Every literal inferred in the 3-valued program completion semantics from $\mathcal{P}$ is also inferred in the 2-valued program completion semantics from $\mathcal{P}$.*

(4) *If $\mathbf{F}_{\mathcal{P}}$ is total, then the 2-valued program completion semantics and the 3-valued program completion semantics make the same inferences from $\mathcal{P}$.*

Exercise 5

(1) Prove Theorem 6.23.
(2) Prove Theorem 6.26.

---

[4]This notation consistently extends the notation of Definition 2.6. Since we are working in a simulation of a 3-valued semantics, our $\mathbf{T}_{\mathcal{P}_3}$ construction is completely analogous to Fitting's original $\Phi_{\mathcal{P}}$ construction, which has a somewhat different definition.

6.6 The stable and well-founded semantics

The stable and well-founded semantics are rather similar to the program completion semantics but capture an intuition that the program completion semantics do not: that a ground atom may be inferred if and only if it is provable in a finite number of steps from negated atoms which are true – or believed – or derived. The construction starts from the ground instantiation of a logic program.

Definition 6.27 *Let P be a normal logic program, and let P^g be its ground instantiation.*

A rule resolution of rules ρ by rule σ is a derivation resolving the head of σ with a positive subgoal of ρ, as below; ρ' is the resolvant of ρ by σ.

$$
\begin{array}{rcl}
\rho & = & a \leftarrow b, c, d, \neg e, \neg f \\
\sigma & = & c \leftarrow d, g, h, \neg i, \neg j \\
\hline
\rho' & = & a \leftarrow b, d, g, h, \neg e, \neg f, \neg i, \neg j.
\end{array}
$$

(Since we treat the body as a set of subgoals, we may list the positive subgoals of the resolvant first and eliminate the duplicate subgoal.)

A transitive rule of P is a rule ρ_0 derivable from ground rules of P^g using 0 or more applications of resolution, as in figure 2, where ρ_n and each σ_i is in P and, for $i < n$, ρ_i is the resolvant of ρ_{i+1} by σ_{i+1}. Here n is called the length of the resolution derivation of ρ_0. The degenerate case, $n = 0$, is that $\rho_0 \in P$.

Thus each rule of P is a transitive rule of P, and if ρ is a transitive rule of P and σ is a rule of P, the resolvant of ρ by σ is also a transitive rule of P.

The program P^* is the set of transitive rules of P with no positive subgoals.

Example 6.28

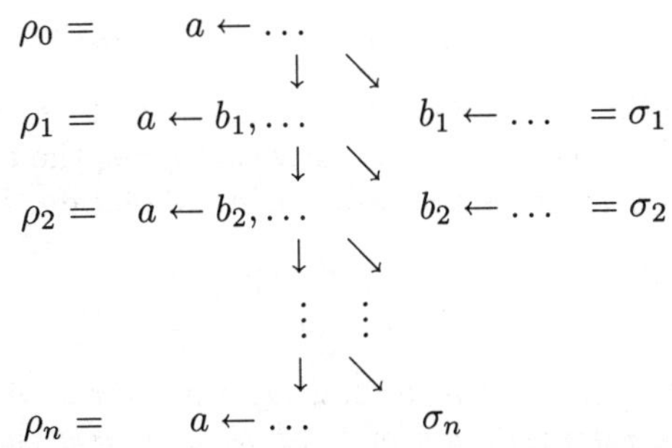

$$
\begin{array}{rl}
\rho_0 = & a \leftarrow \ldots \\
 & \downarrow \searrow \\
\rho_1 = & a \leftarrow b_1, \ldots \qquad b_1 \leftarrow \ldots = \sigma_1 \\
 & \downarrow \searrow \\
\rho_2 = & a \leftarrow b_2, \ldots \qquad b_2 \leftarrow \ldots = \sigma_2 \\
 & \downarrow \searrow \\
 & \vdots \quad \vdots \\
 & \downarrow \searrow \\
\rho_n = & a \leftarrow \ldots \qquad \sigma_n
\end{array}
$$

FIG. 2. Derivation of a transitive rule

(1) Let $\mathcal{P} = \{p \leftarrow q, r, \; q \leftarrow p, r, \; r \leftarrow p, q\}$. Since $\mathcal{P}$ is propositional, $\mathcal{P}^g = \mathcal{P}$. Using one resolution step one can derive the following additional transitive rules:

$$p \leftarrow p, r, \quad p \leftarrow p, q, \quad q \leftarrow p, q, \quad q \leftarrow q, r$$
$$r \leftarrow p, r, \quad r \leftarrow q, r.$$

One more resolution step yields all the remaining transitive rules:

$$p \leftarrow p, q, r, \quad q \leftarrow p, q, r, \quad r \leftarrow p, q, r.$$

Since every transitive rule has a positive subgoal, $\mathcal{P}^* = \emptyset$.

(2) Let $\mathcal{P} = \{p \leftarrow q, r, \; q \leftarrow p, r, \; r \leftarrow p, q, \; r \leftarrow \neg p, \; q \leftarrow \neg r\}$.
As above, from the first three rules alone one can derive the following additional transitive rules:

$$p \leftarrow p, r, \quad p \leftarrow p, q, \quad q \leftarrow p, q, \quad q \leftarrow q, r,$$
$$r \leftarrow p, r, \quad r \leftarrow q, r,$$
$$p \leftarrow p, q, r, \quad q \leftarrow p, q, r, \quad r \leftarrow p, q, r.$$

Now, using the last two rules from $\mathcal{P}$, one can also derive the following transitive rules in one step from the rules of $\mathcal{P}$,

$$p \leftarrow r, \neg r, \quad p \leftarrow q, \neg p, \quad p \leftarrow \neg p, \neg r,$$
$$q \leftarrow p, \neg p, \quad r \leftarrow p, \neg r$$

and the following rules from previously listed transitive rules of $\mathcal{P}$:

$$p \leftarrow p, \neg p, \quad p \leftarrow p, \neg r,$$
$$q \leftarrow p, \neg r, \quad q \leftarrow r, \neg r, \quad q \leftarrow q, \neg p, \quad q \leftarrow \neg p, \neg r,$$
$$r \leftarrow p, \neg p, \quad r \leftarrow r, \neg r, \quad r \leftarrow q, \neg p, \quad r \leftarrow \neg p, \neg r,$$
$$p \leftarrow p, r, \neg r, \quad p \leftarrow p, q, \neg p, \quad p \leftarrow p, \neg p, \neg r,$$
$$q \leftarrow p, r, \neg r, \quad q \leftarrow p, q, \neg p, \quad q \leftarrow p, \neg p, \neg r,$$
$$r \leftarrow p, r, \neg r, \quad r \leftarrow p, q, \neg p, \quad r \leftarrow p, \neg p, \neg r.$$

These turn out to be all the transitive rules of $\mathcal{P}$. Finally, $\mathcal{P}^*$ is

$$\{p \leftarrow \neg p, \neg r, \quad r \leftarrow \neg p, \quad q \leftarrow \neg r, \quad q \leftarrow \neg p, \neg r, \quad r \leftarrow \neg p, \neg r\}.$$

(3) Let $\mathcal{P} = \{q(0) \leftarrow, \; q(s(s(X))) \leftarrow q(X), \; q(X) \leftarrow q(s(s(X)))\}$. In this case, since there are no negative subgoals, the rules in $\mathcal{P}^*$ correspond exactly to literals provable using resolution. So $\mathcal{P}^*$ is

$$\{\; q(0) \leftarrow, \qquad q(s(s(0))) \leftarrow,$$
$$q(s(s(s(s(0))))) \leftarrow, \quad q(s(s(s(s(s(s(0))))))) \leftarrow, \quad \ldots\}.$$

Definition 6.29 *Let $\mathcal{P}$ be a normal logic program. A stable model of $\mathcal{P}$ is an Herbrand interpretation of $\mathcal{P}$ which is a fixed point of $\mathbf{T}_{\mathcal{P}^*}$. The stable semantics infers all sentences (or all ground literals) true in all stable models of $\mathcal{P}$.*

Thus the stable semantics treats the stable models as the intended models of a normal logic program.

Example 6.30 Return to the previous example.
(1) $\mathcal{P} = \{p \leftarrow q, r,\ q \leftarrow p, r,\ r \leftarrow p, q\}$. Since $\mathcal{P}^* = \emptyset$, the only stable model is $\emptyset$, and the stable semantics infers $\neg p, \neg q, \neg r$. (Contrast this with the inferences of the two program completion semantics.)
(2) For $\mathcal{P} = \{p \leftarrow q, r,\ q \leftarrow p, r,\ r \leftarrow p, q,\ r \leftarrow \neg p,\ q \leftarrow \neg r\}$, $\mathcal{P}^*$ is

$$\{p \leftarrow \neg p, \neg r,\ r \leftarrow \neg p,\ q \leftarrow \neg r,\ q \leftarrow \neg p, \neg r,\ r \leftarrow \neg p, \neg r\}.$$

The completion of $\mathcal{P}^*$ is clearly equivalent to

$$\{p \leftrightarrow \neg p \wedge \neg r,\ r \leftrightarrow \neg p,\ q \leftrightarrow \neg r\}.$$

By the first equivalence, p must be false and r must be true. Hence, by the last equivalence, q must be false. And it is easy to check that $\{q\}$ is a model of the completion. Hence the stable semantics infers r, $\neg p$, and $\neg q$. (Again, contrast this with the inferences of the two program completion semantics.)
(3) For $\mathcal{P} = \{q(0) \leftarrow,\ q(s(s(X))) \leftarrow q(X),\ q(X) \leftarrow q(s(s(X)))\}$, $\mathcal{P}^*$ is

$$\{\ q(0) \leftarrow,\qquad q(s(s(0))) \leftarrow,$$
$$q(s(s(s(s(0))))) \leftarrow,\ q(s(s(s(s(s(s(0))))))) \leftarrow,\ \ldots\}$$

and the unique stable model is

$$\{\ q(0),\qquad q(s(s(0))),$$
$$q(s(s(s(s(0))))),\ q(s(s(s(s(s(s(0))))))),\ \ldots\},$$

and the stable semantics infers $q(s^{2n}(0))$ and $\neg q(s^{2n+1}(0))$, for each natural number n. (Again, contrast this with the inferences of the two program completion semantics.)

Example 6.31 Consider again Example 6.3.
(1) Let $\mathcal{P} = \{p \leftarrow p\}$. Every application of resolution just recreates the rule $p \leftarrow p$, so $\mathcal{P}^* = \emptyset$. Thus $\emptyset$ is the only fixed point of $\mathbf{T}_{\mathcal{P}^*}$, i.e., the only stable model of $\mathcal{P}$. So the stable semantics infers $\neg p$ – like the van Emden Kowalski, minimal model, and perfect model semantics, and unlike either program completion semantics. Thus the 2-valued program completion semantics and the stable semantics treat positive recursion differently.
(2) Let $\mathcal{P} = \{a \leftarrow \neg b\}$. Then no rule has any positive subgoals, so no resolution can be applied, and $\mathcal{P}^* = \mathcal{P}$. Thus there is one stable model of $\mathcal{P}$, the unique fixed point of $\mathbf{T}_\mathcal{P}$, which is $\{a\}$. Like all the other semantics discussed here except the minimal model semantics, the stable semantics infers a and $\neg b$.

(3) Let $\mathcal{P} = \{a \leftarrow \neg b, \; z \leftarrow \neg a\}$. Just as in the previous example, $\mathcal{P}^* = \mathcal{P}$, so the stable semantics, like the 2-valued program completion semantics, infers a, $\neg b$, and $\neg z$.

(4) For $\mathcal{P} = \{a \leftarrow \neg b, \; b \leftarrow \neg a, \; c \leftarrow a, \; c \leftarrow b\}$,
$$\mathcal{P}^* \;=\; \{a \leftarrow \neg b, \; b \leftarrow \neg a, \; c \leftarrow \neg b, \; c \leftarrow \neg a\}.$$

The completion of $\mathcal{P}^*$ is equivalent to $\{a \underline{\vee} b, \; c \leftrightarrow \neg a \vee \neg b\}$, from which only the literal c follows – the same conclusions as in the minimal model and 2-valued program completion semantics.

(5) For $\mathcal{P} = \{a \leftarrow \neg a, \; b \leftarrow\}$, $\mathcal{P}^* = \mathcal{P}$ (there being no positive subgoals to resolve on), so the stable semantics makes the contradictory inferences $a, \neg a, b, \neg b$, just as does the 2-valued program completion semantics. The 2-valued program completion semantics and the stable semantics treat negative recursion similarly.

The following lemma shows that the notion of a transitive rule is reasonably robust:

Lemma 6.32 *Let $\mathcal{P}$ be a normal logic program. If*
$$\rho_0 \;=\; a \leftarrow b_1, \ldots, b_k, \neg c_1, \ldots, \neg c_l$$
is the resolvant of ρ_1 by ρ_2, where ρ_1, ρ_2 are transitive rules of $\mathcal{P}$, then there is a transitive rule
$$\rho' \;=\; a \leftarrow b_{i_1}, \ldots, b_{i_{k'}}, \neg c_{j_1}, \ldots, \neg c_{j_{l'}}$$
of $\mathcal{P}$ with the same head as ρ_0 and where every subgoal of ρ' is also a subgoal of ρ_0.

Proof The proof is by induction on the length n of a resolution derivation of ρ_2. If $n = 0$, then $\rho_2 \in \mathcal{P}$, and ρ_0 is the resolvant of a transitive rule of $\mathcal{P}$ by a rule of $\mathcal{P}$ and hence is a transitive rule of $\mathcal{P}$.

Assume the result for all resolution derivations of length n and assume ρ_2 has a resolution derivation of length $n + 1$. That derivation starts out

$$\begin{array}{ll} \rho_2 = & b_1 \leftarrow \ldots \\ & \downarrow \searrow \\ \rho_3 = & b_1 \leftarrow b_2, \ldots \qquad b_2 \leftarrow \ldots = \sigma_3 \\ & \downarrow \searrow \\ & \vdots \qquad\qquad \vdots \end{array}$$

And suppose that ρ_1 is $a \leftarrow b_1, \ldots$. Note that every subgoal of ρ_1, except for b_1, is a subgoal of ρ_0, and every subgoal of ρ_2 is a subgoal of ρ_0.

By inductive hypothesis (applied to the resolvant of ρ_1 by ρ_3), there is a transitive rule
$$\rho'_1 \;=\; a \leftarrow \ldots$$

of $\mathcal{P}$ with head a and whose body is a subset of the body of ρ_1, less b_1, union the body of ρ_3. Thus every subgoal of ρ_1', except for b_2, is in the body of ρ_0. Moreover, every subgoal of σ_3 is in the body of ρ_0.

Now form the resolvant ρ' of ρ_1' with σ_3. Its head is a, and every subgoal is in the body of ρ_1' less b_2 union the body of σ_3, so every subgoal is in the body of ρ_0. Hence ρ' is a transitive rule as desired.

(It is possible that b_2 is a subgoal of ρ_1. In the process of the resolution above, b_2 is removed as a subgoal from ρ', whereas b_2 would remain as a subgoal of ρ_0. Hence ρ_0 and ρ' need not have exactly the same subgoals.)
∎

Exercise 6. The original definition (Gelfond and Lifschitz 1988) of the term *stable model* was the following. Let $\mathcal{P}$ be a ground logic program and $\mathcal{M}$ be an Herbrand interpretation for $\mathcal{P}$.

(1) Form the *Gelfond Lifschitz transformation* $GL(\mathcal{P}, \mathcal{M})$ of $\mathcal{P}$ over $\mathcal{M}$ as follows: (1) Delete from $\mathcal{P}$ each rule with a negative subgoal $\neg a$ where $a \in \mathcal{M}$. (2) Delete all negative subgoals from all remaining rules of $\mathcal{P}$.

(2) Observe that $GL(\mathcal{P}, \mathcal{M})$ is a definite program and thus has a minimum model $\mathcal{N}$. The interpretation $\mathcal{M}$ is a *stable model* of $\mathcal{P}$ if $\mathcal{M} = \mathcal{N}$.

Work through all parts of Example 6.31 using this definition of the term *stable model*. Prove that the two definitions of the term *stable model* are equivalent.

At this point the reader should identify all stable models for the examples in Section 5.

Exercise 7. Let $\mathcal{P}$ be a normal logic program, and let $\mathcal{M}$ be a stable model of $\mathcal{P}$. Prove that $\mathcal{M}$ is also a minimal model of $\mathcal{P}$.

Exercise 8. Find a propositional logic program $\mathcal{P}$ where (1) the completion of $\mathcal{P}$ is consistent (i.e., the 2-valued program completion semantics makes consistent inferences) but where (2) $\mathcal{P}$ has no stable model (i.e., the stable semantics makes contradictory inferences).

****

Proposition 6.33 *For any normal logic program* $\mathcal{P}$, $(\mathcal{P}_3)^* = (\mathcal{P}^*)_3$.

Definition 6.34 *The well-founded semantics infers all ground literals of a normal logic program* $\mathcal{P}$ *which are true in all Herbrand interpretations of* $\mathcal{P}_3$ *which are fixed points of* $\mathbf{T}_{\mathcal{P}_3^*}$. $\mathbf{WF}_\mathcal{P}$ *is the* $\sqsubseteq$-*minimum interpretation* for $\mathcal{P}_3$ which is a fixed point of $\mathcal{P}_3{}^*$.

The results of Theorems 6.23 and 6.26 extend immediately to the well-founded semantics and to the relationship between the well-founded semantics and the stable semantics. In particular, like $\mathbf{F}_{\mathcal{P}}$, $\mathbf{WF}_{\mathcal{P}}$ can be constructed by transfinite induction. In the original definition of the well-founded semantics, the analogue of $\mathbf{WF}_{\mathcal{P}}$ (using a different way of representing partial information) is called the *well-founded partial model of* $\mathcal{P}$.

Example 6.35 Consider again Example 6.3.

(1) Let $\mathcal{P} = \{p \leftarrow p\}$. Since $\mathcal{P}^* = \emptyset$, the well-founded semantics, like the stable semantics (and hence also like the van Emden Kowalski, minimal model, and perfect model semantics, and unlike either program completion semantics), infers $\neg p$. The 3-valued program completion semantics and the well-founded semantics treat positive recursion differently; the stable and well-founded semantics treat positive recursion similarly.

(2) Let $\mathcal{P} = \{a \leftarrow \neg b\}$. Then no rule has any positive subgoals, so no resolution can be applied, and $\mathcal{P}^* = \mathcal{P}$. Hence $\mathcal{P}_3{}^* = \mathcal{P}_3$, and the well-founded semantics, like the 3-valued program completion semantics, infers a and $\neg b$.

(3) $\mathcal{P} = \{a \leftarrow \neg b, \; z \leftarrow \neg a\}$. Since $\mathcal{P}^* = \mathcal{P}$, the well-founded semantics, like the 3-valued program completion semantics, infers a, $\neg b$, and $\neg z$.

(4) For $\mathcal{P} = \{a \leftarrow \neg b, \; b \leftarrow \neg a, \; c \leftarrow a, \; c \leftarrow b\}$,

$$\mathcal{P}^* = \{ \; a \leftarrow \neg b, \quad b \leftarrow \neg a, \quad c \leftarrow \neg b, \quad c \leftarrow \neg a \; \},$$
and
$$\mathcal{P}_3{}^* = \{ \; a \leftarrow \neg b_?, \quad a_? \leftarrow \neg b, \quad b \leftarrow \neg a_?, \quad b_? \leftarrow \neg a,$$
$$c \leftarrow \neg b_?, \quad c_? \leftarrow \neg b, \quad c \leftarrow \neg a_?, \quad c_? \leftarrow \neg a \; \}.$$

Here, as with the 3-valued program completion semantics, both $\{a, b, c\}$ and $\{a_?, b_?, c_?\}$ are fixed points of $\mathbf{T}_{\mathcal{P}_3{}^*}$. So the well-founded semantics also infers no literals. Like the 3-valued program completion semantics, it does not do the "reasoning by cases" that makes the 2-valued semantics infer c.

(5) For $\mathcal{P} = \{a \leftarrow \neg a, \; b \leftarrow\}$, from the fact that $\mathcal{P}^* = \mathcal{P}$, it follows that $\mathcal{P}_3{}^* = \mathcal{P}_3$, and the well-founded semantics infers b – but neither a nor $\neg a$ – just as does the 3-valued program completion semantics. The well-founded semantics treats negative recursion similarly to the 3-valued program completion semantics, and differently from the stable and 2-valued program completion semantics.

Theorem 6.36 *Let $\mathcal{P}$ be a normal logic program, and let $\mathcal{M}$ be an Herbrand interpretation for $\mathcal{P}$. If $\mathcal{M}$ is a fixed point of $\mathbf{T}_{\mathcal{P}^*}$, it is a fixed point* of $\mathbf{T}_\mathcal{P}$. If a literal λ is inferred from $\mathcal{P}$ in the 2-valued program completion semantics, it is also inferred from $\mathcal{P}$ in the stable semantics.

Proof Suppose $\mathcal{M} = \mathbf{T}_{\mathcal{P}^*}(\mathcal{M})$. We must prove that (i) $\mathcal{M} \subseteq \mathbf{T}_\mathcal{P}(\mathcal{M})$, and (ii) $\mathbf{T}_\mathcal{P}(\mathcal{M}) \subseteq \mathcal{M}$.

For (i): Let $a \in \mathcal{M}$; we must show $a \in \mathbf{T}_\mathcal{P}(\mathcal{M})$. Since $\mathcal{M} = \mathbf{T}_{\mathcal{P}^*}(\mathcal{M})$, there is a transitive rule

$$\rho_0 \;=\; a \leftarrow \neg c_1, \ldots, \neg c_m$$

of $\mathcal{P}$, where each c_i is a ground atom of $\mathcal{P}$ and no $c_i \in \mathcal{M}$.

Consider a derivation for ρ_0, as in figure 3. Since ρ_0 has no positive subgoals, for each of the d_j's, some subderivation of the above must be a derivation of a rule

$$d_j \leftarrow \neg c_{l_1^j}, \ldots, \neg c_{l_{f_j}^j}$$

of $\mathcal{P}^*$; since $\mathcal{M} = \mathbf{T}_{\mathcal{P}^*}(\mathcal{M})$, each $d_j \in \mathcal{M}$. Thus, all the subgoals of rule ρ_n are true in $\mathcal{M}$, so $a \in \mathbf{T}_\mathcal{P}(\mathcal{M})$.

Part (ii) is similar and is left for the reader. The last statement of the theorem is now immediate from the definitions. ∎

It follows that:

Theorem 6.37 *Let $\mathcal{M}$ be an Herbrand interpretation for $\mathcal{P}_3$. If $\mathcal{M}$ is a fixed point of $\mathbf{T}_{\mathcal{P}_3^*}$, it is a fixed point of $\mathbf{T}_{\mathcal{P}_3}$. If a literal λ is inferred from $\mathcal{P}$ in the 3-valued program completion semantics, it is also inferred from $\mathcal{P}$ in the well-founded semantics.*

Proposition 6.38 *Suppose no rule in program $\mathcal{P}$ has any positive subgoals. Then:*

- *$\mathcal{P}^* = \mathcal{P}$ and $\mathcal{P}_3^* = \mathcal{P}_3$*
- *The stable semantics makes the same inferences from $\mathcal{P}$ as does the 2-valued program completion semantics.*

$$\begin{array}{ccc}
\rho_0 = & a \leftarrow \ldots & \\
& \downarrow \quad \searrow & \\
\rho_1 = & a \leftarrow b_1, \ldots & \quad b_1 \leftarrow \ldots = \sigma_1 \\
& \downarrow \quad \searrow & \\
\rho_2 = & a \leftarrow b_2, \ldots & \quad b_2 \leftarrow \ldots = \sigma_2 \\
& \downarrow \quad \searrow & \\
& \vdots \quad \vdots & \\
& \downarrow \quad \searrow & \\
\rho_n = & a \leftarrow d_1, \ldots, d_g, \neg c_{i_1}, \ldots, \neg c_{i_h} & \quad \sigma_n
\end{array}$$

FIG. 3. Derivation of ρ_0

- *The well-founded semantics makes the same inferences from $\mathcal{P}$ as does the 3-valued program completion semantics.*

The original definition of the well-founded semantics used a different construction.[5]

Definition 6.39 *Let $\mathcal{P}$ be a normal logic program, let $\mathcal{M}$ be an Herbrand interpretation for $\mathcal{P}_3$, and let U be a set of ground atoms $a_?$ of $\mathcal{P}_3$. Then U is unfounded over $\mathcal{M}$ if, for every $a_? \in U$ and every rule*

$$a_? \leftarrow b_{1?}, \ldots, b_{k?}, \neg c_1, \ldots, \neg c_m$$

of $\mathcal{P}_3$, one (or both) of the following is true:

- *some $b_{i?} \notin \mathcal{M}$ or some $c_j \in \mathcal{M}$, or*
- *some $b_{i?} \in U$.*

That is, U is disjoint from $\mathbf{T}_{\mathcal{P}_3}(\mathcal{M} - U)$.

Theorem 6.40 $\mathbf{WF}_\mathcal{P}$ *is the $\sqsubseteq$-minimum fixpoint $\mathcal{M}$ of $\mathbf{T}_{\mathcal{P}_3}$, where, for all sets U of ground atoms $a_?$ of $\mathcal{P}_3$, if U is unfounded over $\mathcal{M}$, then $U \cap \mathcal{M} = \emptyset$.*

Exercise 9. Prove Theorem 6.40. (Suggestion: show that if $\mathcal{M} = \mathbf{T}_{\mathcal{P}_3}(\mathcal{M})$ and $U \cap \mathcal{M} = \emptyset$ for all unfounded U, then $\mathbf{T}_{\mathcal{P}_3{}^*}(\mathcal{M}) \sqsubseteq \mathcal{M}$.)

Exercise 10

(1) Let $\mathcal{P}$ be a normal logic program and suppose $\mathbf{WF}_\mathcal{P}$ is total. Prove that the stable and well-founded semantics infer the same literal from $\mathcal{P}$.

(2) Let $\mathcal{P}$ be a normal logic program and suppose that $\mathbf{F}_\mathcal{P}$ is total. Prove that both program completions semantics, the well-founded semantics, and the stable semantics all infer the same literals from $\mathcal{P}$.

(3) Let $\mathcal{P}$ be a stratified normal logic program, and let $\mathcal{M}$ be its perfect model. Prove that $\mathbf{WF}_\mathcal{P} = \widehat{\mathcal{M}}$. Infer that for $\mathcal{P}$ a stratified program, the well-founded and stable semantics infer the same literals as does the perfect model semantics.

Exercise 11. Let $\mathcal{P}$ be a normal logic program. Prove that $\mathcal{P} \cup \mathbf{WF}_\mathcal{P}$ has a minimal model, and that every minimal model of $\mathcal{P} \cup \mathbf{WF}_\mathcal{P}$ is also a minimal model of $\mathcal{P}$.

---

[5]The definition of the well-founded semantics presented above is modified from (Schlipf 1990, 1992). It inspired in part by several different constructions of the well-founded semantics, notably (Van Gelder 1989b; Przymusiński 1989; Ross 1989). Przymusiński (1989) was the first to realize that the well-founded and stable semantics are just 3-valued and 2-valued versions of each other; Dung (1992) gave another development of this equivalence.

Both the stable and well-founded semantics have proved to be robust concepts, having a variety of equivalent constructions. There are second order logic characterizations by Van Gelder (Van Gelder and Schlipf 1993) and by Dung (1992). There are extensive developments of these and related semantics in terms of negations as hypotheses, argumentation, and relationships with abductive inference; this has been developed in several papers, e.g., (Eshghi and Kowalski 1989; Dung 1991; Kakas and Mancarella 1991; Bondarenko et al 1993). In the latter approach to semantics, negative literals (basically, the literals $\neg a_?$ of our treatment) are treated as hypotheses that can be made, and the logic program (plus consistency requirements) provides tools for argumentation that allow one set of hypotheses to be used in arguments with another. Based upon the exact form of argumentation used, several different semantics can be developed.

6.7 Summary: a comparison of inferences made

We summarize in figure 4 theorems from earlier in this section comparing the inferences made in the various semantics. We draw an arrow from a semantics S_1 to a semantics S_2 in the chart below if for all programs $\mathcal{P}$, $Inferences_{S_1}(\mathcal{P}) \subseteq Inferences_{S_2}(\mathcal{P})$, where we define the inferences always to be a set of ground literals, not sentences.[6]

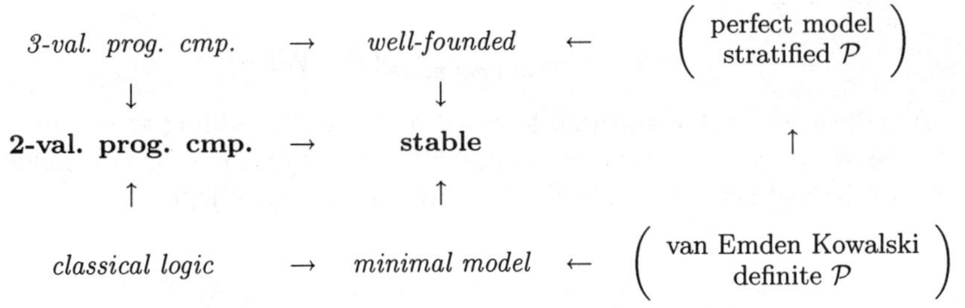

$$
\begin{array}{ccccc}
\textit{3-val. prog. cmp.} & \rightarrow & \textit{well-founded} & \leftarrow & \left(\begin{array}{c} \text{perfect model} \\ \text{stratified } \mathcal{P} \end{array} \right) \\
\downarrow & & \downarrow & & \\
\textbf{2-val. prog. cmp.} & \rightarrow & \textbf{stable} & & \uparrow \\
\uparrow & & \uparrow & & \\
\textit{classical logic} & \rightarrow & \textit{minimal model} & \leftarrow & \left(\begin{array}{c} \text{van Emden Kowalski} \\ \text{definite } \mathcal{P} \end{array} \right)
\end{array}
$$

FIG. 4. A comparison of inferences made

We shall write two semantics in parentheses to indicate that they do not apply to all logic programs. We shall write semantics which make consistent inferences on all normal logic programs in italic font, and semantics which make inconsistent inferences on some logic programs in bold face font.

Exercise 12. Modify Figure 4 to indicate the inferences made only on stratified logic programs.

---

[6] Recall that we did not define which sentences are inferred in the 3-valued semantics, so we could not adequately compare, say, the stratified semantics and the well-founded semantics if we compared two semantics based upon the sentences inferred.

6.8 Other approaches

Researchers in logic programming semantics continue to work in a variety of areas. Reasonable negation-as-failure semantics are becoming available for larger classes of logic programs, notably logic programs with disjunctions of atoms or negative literals in their heads. Also, other semantics have been suggested for normal logic programs. One direction has been to look for natural semantics that are stronger than the well-founded semantics but weaker than the stable. We mention here some of the suggestions that have been made:

(1) Infer a if a is true in all $\sqsubseteq$-*maximal consistent* fixed points of $\mathbf{T}_{\mathcal{P}_3{}^*}$; infer $\neg a$ if $a_?$ is false in all such fixed points.

(2) Strengthen the well-founded semantics by finding more ways to add negative literals. When the well-founded semantics infers a literal $\neg a$ (by inferring $\neg a_?$), no later inference will force a to be true. Look for larger sets of negative literals with the same property where they can all be consistently inferred.

(3) Blend the well-founded and minimal model semantics. This has been particularly important in the study of logic programs with disjunctions in their heads (see, e.g., (Baral *et al.* 1991)), but it can be used here too, as in the semantics (Dix 1990) which makes the following inferences:

$$Inferences_{minimal\ model}(\mathcal{P} \cup \mathbf{WF}_\mathcal{P}).$$

(4) Adjust the stable and well-founded semantics by adding to $\mathcal{P}^*$ more rules, more ways to derive positive literals, for example by new kinds of derived rules, to make it a bit more like classical logic:

$$\frac{p \leftarrow a, \neg b, \neg p}{p \leftarrow a, \neg b} \qquad \frac{\begin{array}{l}p \leftarrow a, b, \neg c\\ p \leftarrow d, \neg b, \neg f\end{array}}{p \leftarrow a, d, \neg c, \neg f}$$

But there are limitations; see Exercise 13.

7 Intuitions

7.1 On precedences and the minimal model semantics

Researchers in logic programming have tended to find that the minimal model semantics makes too few inferences. Recall the program $a \leftarrow \neg b$. In classical logic, this is equivalent to $a \vee b$, so the minimal model semantics infers neither a nor $\neg a$, neither b nor $\neg b$. Many have argued that this shows the minimal model semantics to be too weak a form of negation as failure. The minimal model semantics can be thought of as trying to minimize the set of true ground atoms, giving equal priority to minimizing all the atoms. And, to many researchers, the program $\{a \leftarrow \neg b\}$ should be taken as giving unequal priorities to the two atoms. As discussed in the definition of the semantics for stratified programs, the intuition is that, since b appears only in the rule defining a, the semantics should put a higher priority upon "making b false" than upon "making a false." Is this prioritization reasonable? Note that if this prioritization is unreasonable, then the later semantics we have discussed are all probably also unreasonable.

7.2 Intuitions: 3-valued and 2-valued

For non-stratified normal logic programs, especially when $\mathbf{WF}_\mathcal{P}$ is not total, there is in general no agreement upon the proper set of inferences. At this point it often seems that different semantics have different intuitions, and that the semantics must be chosen to match the circumstances.

We have considered two semantics based upon 3-valued (or 4-valued) logic, the 3-valued program completion semantics and the well-founded semantics. The remaining semantics we have considered are based upon 2-valued logics. In particular, the 2-valued and 3-valued program completion semantics are 2- and 3-valued versions of each other. Similarly, the stable and well-founded semantics are 2- and 3-valued versions of each other. Different intuitions and interpretations seem to lead to the 2-valued and the 3-valued semantics.

Example 7.1 Consider yet again Example 6.3(5).
$$\mathcal{P} = \{a \leftarrow \neg a, \ b \leftarrow \}.$$
$$\mathcal{P}_3 = \{a \leftarrow \neg a_?, \ a_? \leftarrow \neg a, \ b \leftarrow, \ b_? \leftarrow \}.$$

There are no fixed points of $\mathbf{T}_\mathcal{P}$, so the 2-valued program completion semantics makes inconsistent inferences. On the other hand, there are two fixed points, of $\mathbf{T}_{\mathcal{P}_3}$, $\{a, b, b_?\}$ and $\{a_?, b, b_?\}$, so the 3-valued version infers b.

One way to look at the difference between the 2-valued and 3-valued approaches is to think of the 2-valued semantics as providing methods of specification and the 3-valued as providing methods of making inferences. The 2-valued semantics can be thought of as treating $\mathcal{P}$ as an inconsistent specification or definition, which is thus nonsense. If the specification is inconsistent, it should be rewritten.

V. S. Subrahmanian justifies 3-valued treatments on the ground that such a program $\mathcal{P}$ might contain information gathered from several experts. The experts have provided contradictory information regarding a. That is not surprising; experts disagree. Our job is to infer as much as we reasonably can from admittedly imperfect data. We see that our assumptions contain contradictory information about a, so we infer nothing at all. But we see that we still have good grounds for inferring b.

The 2-valued semantics supporter might respond that this is a distortion of the purpose of a logic. The logic should tell us that something is inconsistent, but it should not try to rescue us by providing dubious information. The supporter of 3-valued semantics would reply that it is very well to document inconsistencies, but we may never eliminate them in, for example, an expert system; the function of an expert system is to provide reasonable inferences, not to document that the experts are imperfect.

Listed below are a set of intuitions which might lead to 2-valued or 3-valued semantics. Of course, this is our list, and other authors will have somewhat different opinions.

(1) **The 2-valued semantics:**
- Are defined in terms of ordinary 2-valued models, and seem often more natural when a program is used to specify the behavior of a system or to describe causal relationships.
- Allow arguments by breaking into cases, as with the program

$$\{a \leftarrow \neg b, \quad b \leftarrow \neg a, \quad c \leftarrow a, \quad c \leftarrow \neg a\}.$$

From this program, the 2-valued program completion and stable semantics infer c; the 3-valued program completion and well-founded semantics infer no ground literals.
- Allow for abductive inference – reasoning from effect to cause. (See Example 5.2(6).)
- Coincide more directly with other non-monotonic logic formalisms, e.g., Reiter's default logic and Moore's autoepistemic logic.
- Are more expressive than the 3-valued semantics. (See the last item under the 3-valued semantics below.)

(2) **The 3-valued semantics:**
- Are defined in terms of 3-valued logic, which corresponds naturally to human beliefs – we may believe a proposition is true, or believe it is false, or have no opinion – and perhaps occasionally we may believe it both true and false.
- Always make consistent, albeit sometimes partial, sets of inferences – allowing inferences such as from the experts of the preceding example.
- Inferences must be drawn in a "constructive" logic, and proofs are uniform. There are no proofs by contradiction or other even marginally suspect proof methods. Some feel that the uses of the 2-valued semantics for abduction are examples of proof by a sort of contradiction.
- Seem more natural when negation as failure is treated as an heuristic for inferring what the programmer is likely to have intended. Humans seem very often to use "if" in a very constructive manner, and that expectation is one way to justify many

of the negation as failure techniques discussed here. The non-constructive aspects of the 2-valued versions discussed above may suggest that the heuristic is being pushed too far.
- Logic programming semantics are sometimes evaluated with standards appropriate to programming language semantics. The 3-valued semantics are more computationally tractable than the 2-valued semantics.

  In particular, when a fixed, function-free, normal logic program is used to draw inferences over varying databases of information, making inferences in the 3-valued program completion and well-founded semantics can be done in polynomial time (as a function of the size of the databases), while for the 2-valued program completion and stable semantics making inferences is $\mathcal{NP}$-complete.

  Of course, what this really says is that the 2-valued semantics have greater expressive power than the 3-valued semantics over finite databases D. The 3-valued semantics can express only polynomial-time-computable queries, while the 2-valued semantics can describe $\mathcal{NP}$ queries, such as the existence of a Hamiltonian cycle in a graph or the existence of a satisfying assignment for a propositional formula.

Exercise 13. Let $\mathcal{P}$ be a normal logic program. Recall that $\mathcal{P}^*$ is the set of all rules

$$a \leftarrow \neg c_1, \ldots, \neg c_k$$

where $c_1, \ldots, c_k$ are ground atoms of $\mathcal{P}$ and the rule can be derived from ground of instances rules of $\mathcal{P}$ using a special form of resolution. By comparison, define $\mathcal{P}^{\ddagger}$ to be the set of all rules

$$a \leftarrow \neg c_1, \ldots, \neg c_k$$

where $a \vee c_1 \vee \cdots \vee c_k$ is a logical consequence of $\mathcal{P}$ in first order logic. Using $\mathcal{P}^{\ddagger}$, define semantics for $\mathcal{P}$ analogously to the way the stable and well-founded semantics were defined using $\mathcal{P}^*$. What can you say about those semantics?

Exercise 14

(1) Find an algorithm which, given a propositional logic program $\mathcal{P}$, constructs $\mathbf{F}_\mathcal{P}$ in time polynomial in the size of $\mathcal{P}$.

(2) Find an algorithm which, given a propositional logic program $\mathcal{P}$, constructs $\mathbf{WF}_\mathcal{P}$ in time polynomial in the size of $\mathcal{P}$.

(3) (Kolaitis and Papadimitriou 1991) Show that checking whether a propositional normal logic program $\mathcal{P}$ has a model of its completion

is $\mathcal{NP}$-complete.
Suggested approach: Take an instance of the conjunctive normal form (CNF) satisfiability problem, e.g.,

$$((p \vee q \vee \neg r) \wedge (\neg p \vee q) \wedge (\neg q \vee r)).$$

Part of the construction is to translate this formula into a series of atoms – atoms which describe the syntax – e.g.

isPositiveDisjunct(p, c_1)
isNegativeDisjunct(q, c_3)

where c_i represents the i'th clause in the CNF formula. Now add extra rules so that the fixed points of $\mathbf{T}_\mathcal{P}$ correspond 1-1 to the satisfying interpretations for the original formula. (We are actually describing a first-order logic program with no function symbols, but its ground instantiation is then equivalent to a finite propositional logic program.) Look at Example 5.2(6) for techniques.

(4) (Marek and Truszczyński 1991) Show that checking whether a propositional normal logic program $\mathcal{P}$ has a stable model is $\mathcal{NP}$-complete.

8 Formalisms for comparing semantics

Many different intuitions seem to have led people to prefer one logic programming semantics to another. Here we summarize some of the differences among the semantics that people have remarked upon. Whether or not they are used for evaluation, they represent important points for comparison. We start with some simple criteria. The reader should notice which of the semantics discussed in this chapter meet each criterion.

(1) **Divergences from classical logic:**
- Is *Inferences*$(\mathcal{P})$ always consistent? (Recall that, treated as a set of implications in classical logic, every logic program has a model.)
- Does *Inferences*$(\mathcal{P})$ contain all classical consequences of $\mathcal{P}$?
- Does *Inferences*$(\mathcal{P})$ contain all classical consequences of $\mathcal{P}\, \cup$ *Inferences*$(\mathcal{P})$?
- Is there an underlying notion of a 2-valued model of $\mathcal{P}$, where the set of inferences is the set of all sentences or literals true in all such models?

(2) **Comparison with the van Emden Kowalski prototype:**
For definite programs $\mathcal{P}$, does the semantics make the same inferences as the van Emden Kowalski semantics does?

(3) **Divergences from the minimal model semantics:**
- Is *Inferences*$(\mathcal{P})\, \cup$ *Inferences*$_{minimal\ model}(\mathcal{P})$ consistent? That is, is there a minimal model $\mathcal{M}$ of $\mathcal{P}$ which is also a model of

Inferences($\mathcal{P}$)?
- Is *Inferences*$_{minimal\ model}$($\mathcal{P}$) $\subseteq$ *Inferences*($\mathcal{P}$)?

(4) **Does the semantics distinguish between a rule and its contrapositive:**
For example, is *Inferences*($\{a \leftarrow \neg b\}$) = *Inferences*($\{b \leftarrow \neg a\}$)?

8.1 Standards of acceptability as a logic

We discuss here three standards suggested by Dix (see, e.g., (Dix 1990)), building upon ideas from (Kraus *et al.* 1990). Dix suggests that a reasonable logic programming semantics should satisfy the following properties. Though many of the properties of classical logic are explicitly rejected in logic programming semantics, Dix suggests that some weaker properties should nevertheless be required. We use a and b below to range over ground atoms of a program $\mathcal{P}$.[7]

For this subsection we identify, as is commonly done, an atom a with the rule $a \leftarrow$.

- **Cut:** If $a \in$ *Inferences*($\mathcal{P}$) then *Inferences*($\mathcal{P} \cup \{a\}$) $\subseteq$ *Inferences*($\mathcal{P}$) (Note that *Inferences*($\mathcal{P} \cup \{a\}$) may be restricted to being a set of ground literals.)
  Cut asserts, essentially, that one can prove and then use lemmas. Here, because of the syntax of normal logic programs, these lemmas are restricted to atoms a.

- **Cumulative (or cautious) monotonicity:** If $a \in$ *Inferences*($\mathcal{P}$) then *Inferences*($\mathcal{P}$) $\subseteq$ *Inferences*($\mathcal{P} \cup \{a\}$)
  As noted earlier, monotonicity is commonly rejected in logic programming. But cumulative monotonicity is a much weaker property. It merely says that nothing is lost in taking as an assumption any atom that has already been proved. Consider cumulative monotonicity from the point of view of practical theorem-proving. Once we prove a statement a, we add a to a list of known facts, rather than having to rederive a each time we need it. But that seems to amount to adding a to our list of assumptions. Cumulative monotonicity states that this is harmless.

- **Rational monotonicity:** If $\neg a \notin$ *Inferences*($\mathcal{P}$) then *Inferences*($\mathcal{P}$) $\subseteq$ *Inferences*($\mathcal{P} \cup \{a\}$).
  Rational monotonicity is a measure of how non-monotonic the inference operator is: here, adding an atom as a new assumption will not lead us to withdraw any inferences unless that atom explicitly contradicts previous inferences.

Exercise 15. Prove that all the semantics discussed in this paper satisfy cut.

---

[7]The results of this subsection are all due to Dix.

Theorem 8.1 *Cumulative and rational monotonicity fail for both the 2-valued program completion semantics and the stable semantics.*

Proof Let

$$\mathcal{P} = \{a \leftarrow \neg b, \quad b \leftarrow \neg a, \quad c \leftarrow \neg c, \quad c \leftarrow \neg a\}.$$

The completion of $\mathcal{P}$ is

$$\{a \leftrightarrow \neg b, \quad b \leftrightarrow \neg a, \quad c \leftrightarrow \neg c \vee \neg a\}.$$

If we restrict inferences to literals,

$$\textit{Inferences}_{\textit{2-valued program completion}}(\mathcal{P}) = \{\neg a, b, c\}.$$

By Proposition 6.38, the stable semantics makes the same inferences from $\mathcal{P}$.

Now consider $\mathcal{P} \cup \{c\}$. It is routine to check that $\{a,c\}$ and $\{b,c\}$ are both stable models of $\mathcal{P} \cup \{c\}$, so neither semantics infers either $\neg a$ or b from $\mathcal{P} \cup \{c\}$. ∎

Theorem 8.2 *The 3-valued program completion semantics satisfies both cumulative and rational monotonicity.*

Proof First prove rational monotonicity: Let $\mathcal{P}$ be a normal logic program. Assume $\neg a \notin \textit{Inferences}(\mathcal{P})$, i.e., $a_? \in \mathbf{F}_\mathcal{P}$. We need to show that $\mathbf{F}_\mathcal{P} \sqsubseteq \mathbf{F}_{\mathcal{P} \cup \{a\}}$. We use the fact that, in any lattice $(L, \sqsubseteq)$, $A \sqsubseteq B$ if and only if $A \sqsubseteq A \sqcap B$.

We shall show that

$$\mathbf{T}_{\mathcal{P}_3}(\mathbf{F}_\mathcal{P} \sqcap \mathbf{F}_{\mathcal{P} \cup \{a\}}) \sqsubseteq \mathbf{F}_\mathcal{P} \sqcap \mathbf{F}_{\mathcal{P} \cup \{a\}}.$$

Since $\mathbf{F}_\mathcal{P} = \sqcap \{I : \mathbf{T}_{\mathcal{P}_3}(I) \sqsubseteq I\}$, it follows that $\mathbf{F}_\mathcal{P} \sqsubseteq \mathbf{F}_\mathcal{P} \sqcap \mathbf{F}_{\mathcal{P} \cup \{a\}}$, and hence that $\mathbf{F}_\mathcal{P} \sqsubseteq \mathbf{F}_{\mathcal{P} \cup \{a\}}$.

By monotonicity,

$$\mathbf{T}_{\mathcal{P}_3}(\mathbf{F}_\mathcal{P} \sqcap \mathbf{F}_{\mathcal{P} \cup \{a\}}) \sqsubseteq \mathbf{T}_\mathcal{P}(\mathbf{F}_\mathcal{P}) = \mathbf{F}_\mathcal{P}, \quad \text{and similarly} \quad (8.1)$$

$$\mathbf{T}_{(\mathcal{P} \cup \{a\})_3}(\mathbf{F}_\mathcal{P} \sqcap \mathbf{F}_{\mathcal{P} \cup \{a\}}) \sqsubseteq \mathbf{T}_{(\mathcal{P} \cup \{a\})_3}(\mathbf{F}_{\mathcal{P} \cup \{a\}}) = \mathbf{F}_{\mathcal{P} \cup \{a\}}. \quad (8.2)$$

Note that $(\mathcal{P} \cup \{a\})_3 = \mathcal{P}_3 \cup \{a, a_?\}$, so, for any Herbrand interpretation $\mathcal{M}$ for $\mathcal{P}_3$, $(\mathcal{P} \cup \{a\})_3(\mathcal{M}) = \mathcal{P}_3(\mathcal{M}) \cup \{a, a_?\}$. By (8.1), since $a_? \in \mathbf{F}_\mathcal{P}$, $a_? \in \mathbf{T}_{\mathcal{P}_3}(\mathbf{F}_\mathcal{P} \sqcap \mathbf{F}_{\mathcal{P} \cup \{a\}})$. Hence

$$\begin{aligned}\mathbf{T}_{(\mathcal{P} \cup \{a\})_3}(\mathbf{F}_\mathcal{P} \sqcap \mathbf{F}_{\mathcal{P} \cup \{a\}}) &= \mathbf{T}_{\mathcal{P}_3}(\mathbf{F}_\mathcal{P} \sqcap \mathbf{F}_{\mathcal{P} \cup \{a\}}) \cup \{a, a_?\} & (8.3)\\ &= \mathbf{T}_{\mathcal{P}_3}(\mathbf{F}_\mathcal{P} \sqcap \mathbf{F}_{\mathcal{P} \cup \{a\}}) \cup \{a\}. & (8.4)\end{aligned}$$

So, from (8.4), by definition of $\sqsubseteq$,

$$\begin{aligned}\mathbf{T}_{\mathcal{P}_3}(\mathbf{F}_\mathcal{P} \sqcap \mathbf{F}_{\mathcal{P} \cup \{a\}}) &\sqsubseteq \mathbf{T}_{(\mathcal{P} \cup \{a\})_3}(\mathbf{F}_\mathcal{P} \sqcap \mathbf{F}_{\mathcal{P} \cup \{a\}}) & (8.5)\\ &\sqsubseteq \mathbf{F}_{\mathcal{P} \cup \{a\}}. & (8.6)\end{aligned}$$

By (8.1) and (8.6), $\mathbf{T}_{\mathcal{P}_3}(\mathbf{F}_\mathcal{P} \sqcap \mathbf{F}_{\mathcal{P} \cup \{a\}}) \sqsubseteq \mathbf{F}_\mathcal{P} \sqcap \mathbf{F}_{\mathcal{P} \cup \{a\}}$, as desired.

Now show that the 3-valued program completion semantics satisfies cumulative monotonicity. Suppose that $a \in \mathit{Inferences}(\mathcal{P})$. Then, since the semantics always makes consistent inferences, $\neg a \notin \mathit{Inferences}(\mathcal{P})$. So, by rational monotonicity, $\mathbf{F}_\mathcal{P} \sqsubseteq \mathbf{F}_{\mathcal{P} \cup \{a\}}$, as desired. ∎

Exercise 16. Prove that the well-founded semantics satisfies cumulative and rational monotonicity. (**Suggestion:** First prove the following lemma: If $a_? \in \mathbf{T}_{\mathcal{P}_3^*}(I)$ then $\mathbf{T}_{\mathcal{P}_3^*}(I) \sqsubseteq \mathbf{T}_{(\mathcal{P} \cup \{a\})_3^*}(I)$.)

Exercise 17. Show that the van Emden Kowalski semantics (for definite programs) and the perfect model semantics (for stratified programs) satisfy cumulative and rational monotonicity.

Exercise 18. Does the minimal model semantics satisfy cumulative monotonicity? Rational monotonicity?

8.2 Standards of acceptability for programming

In this section, we shall discuss a standard we have proposed, called the *principle of stratification* due to its relationship to (and inspiration from) the stratified semantics. Related ideas have been proposed by Maher and Dix.

Definition 8.3 *A pair $(\mathcal{P}, \mathcal{Q})$ of normal programs is a stratified pair if, for each relation symbol or proposition letter r, if r appears in the head of a rule in $\mathcal{Q}$, then r does not appear at all in $\mathcal{P}$.*

When we are discussing a stratified pair $(\mathcal{P}, \mathcal{Q})$, we compare the inferences of $\mathcal{P} \cup \mathcal{Q}$ with the inferences of $\mathcal{P}$ alone. For $\mathcal{P}$ alone, we evaluate those inferences over the Herbrand universe for $\mathcal{P} \cup \mathcal{Q}$.

The principle of stratification describes a logic programming semantics; it is a property a semantics may or may not satisfy. The intuition for the principle of stratification (below) is the same as one intuition for the stratified semantics. In a stratified pair $(\mathcal{P}, \mathcal{Q})$, $\mathcal{P}$ should be thought of as defining a set of relations, and $\mathcal{Q}$ should be thought of as using those relations to define other relations.

Definition 8.4 *The principle of stratification asserts that for any stratified pair $(\mathcal{P}, \mathcal{Q})$ of normal logic programs:*

(1) *If relation symbol r appears in any rule in $\mathcal{P}$, then for any term t,*

$$r(t) \in \mathit{Inferences}(\mathcal{P}) \quad \text{if and only if} \quad r(t) \in \mathit{Inferences}(\mathcal{P} \cup \mathcal{Q}).$$

(2) *If $\mathcal{P}$ and $\mathcal{P}'$ use exactly the same constant, function, and relation symbols (and hence $(\mathcal{P}', \mathcal{Q})$ is also a stratified pair), and if*

$$\mathit{Inferences}(\mathcal{P}) = \mathit{Inferences}(\mathcal{P}')$$

then
$$\text{Inferences}(\mathcal{P} \cup \mathcal{Q}) = \text{Inferences}(\mathcal{P}' \cup \mathcal{Q}).$$

Suppose logic programs are used to define sets of inferences. And suppose that two programmers are working together. One programmer writes $\mathcal{P}$ to define certain relations. The other programmer writes $\mathcal{Q}$ to define other relations, using the relations defined in $\mathcal{P}$. The result is a stratified pair $(\mathcal{P}, \mathcal{Q})$. Part (1) of the principle of stratification asserts that module $\mathcal{Q}$ does not affect the definitions made in $\mathcal{P}$. Part (2) asserts a sort of independence of the form of the program, similar to the implementational independence discussed in object-oriented programming: it does not matter how $\mathcal{P}$ is written – it only matters what inferences are made from $\mathcal{P}$.

Theorem 8.5 *Part (1) of the principle of stratification fails for both the 2-valued program completion semantics and the stable semantics. If we define Inferences($\mathcal{P}$) to be a set of ground literals for any logic program $\mathcal{P}$, then Part (2) of the principle of stratification also fails.*

Proof (1) Consider the example
$$\mathcal{P} = \{a \leftarrow \neg b, \ b \leftarrow \neg a\} \qquad \mathcal{Q} = \{c \leftarrow \neg c, \ c \leftarrow \neg a\}.$$

Obviously, $(\mathcal{P}, \mathcal{Q})$ is a stratified pair. Since no rule in either $\mathcal{P}$ or $\mathcal{Q}$ contains positive subgoals, the two semantics make the same inferences from $\mathcal{P}$, and they also make the same inferences from $\mathcal{P} \cup \mathcal{Q}$ (see Proposition 6.38). The completion of $\mathcal{P}$ is
$$\{a \leftrightarrow \neg b, \ b \leftrightarrow \neg a\},$$
which "says" that exactly one of a and b is true. The completion of $\mathcal{P} \cup \mathcal{Q}$ is
$$\{a \leftrightarrow \neg b, \ b \leftrightarrow \neg a, \ c \leftrightarrow \neg c \vee \neg a\}.$$

The last statement is satisfied only with c true and a false. Hence in both semantics
$$\neg a \notin \text{Inferences}(\mathcal{P}) \quad \text{but} \quad \neg a \in \text{Inferences}(\mathcal{P} \cup \mathcal{Q}),$$
contradicting part (1) of the principle of stratification since b appears in $\mathcal{Q}$.

(2) Consider the example
$$\begin{aligned}
\mathcal{P} &= \{a \leftarrow \neg b, \quad b \leftarrow \neg a, \quad c \leftarrow \neg d, \quad d \leftarrow \neg c\} \\
\mathcal{P}' &= \{a \leftarrow \neg c, \quad c \leftarrow \neg a, \quad b \leftarrow \neg d, \quad d \leftarrow \neg b\} \\
\mathcal{Q} &= \{e \leftarrow \neg a, \quad e \leftarrow \neg b\}.
\end{aligned}$$

Again, since all subgoals are negative, the two semantics make the same inferences. The completion of $\mathcal{P}$ is equivalent to $\{a \leftrightarrow \neg b, \ c \leftrightarrow$

$\neg d\}$. The completion of $\mathcal{P}'$ is equivalent to $\{a \leftrightarrow \neg c, \ b \leftrightarrow \neg d\}$. Hence $\textit{Inferences}(\mathcal{P}) = \textit{Inferences}(\mathcal{P}') = \emptyset$.

But the completion of $\mathcal{P} \cup \mathcal{Q}$ is equivalent to

$$\{a \leftrightarrow \neg b, \quad c \leftrightarrow \neg d, \quad e \leftrightarrow \neg a \vee \neg b\},$$

while the completion of $\mathcal{P} \cup \mathcal{P}'$ is equivalent to

$$\{a \leftrightarrow \neg c, \quad b \leftrightarrow \neg d, \quad e \leftrightarrow \neg a \vee \neg b\}.$$

The atom e follows from the first of these, but not from the second, contradicting part (2) of the principle of stratification. ∎

Some researchers use the stable semantics for abductive reasoning, as in Example 5.2(6). Thus they are exploiting the failure of the principle of stratification. The intuitions for such abductive reasoning directly contradict the intuitions motivating the principle of stratification. There is no consensus among logic programmers about which is the proper intuition. In the end, this may be one of the places where people choose a semantics depending upon what they are trying to accomplish.

When inferences are restricted to being ground atoms, the second part of the principle of stratification can be thought of as some sort of constructivity assumption. In the second example above, the completion of $\mathcal{P}$ implied $a \vee b$ without implying anything about the truth of a or b. In a semantics (such as the 3-valued program completion semantics) which satisfies the principle of stratification, there is no way in which a program can imply such a disjunction without implying one of the disjuncts.

Theorem 8.6 *The 3-valued program completion semantics satisfies the principle of stratification.*

Proof (1) Let $(\mathcal{P}, \mathcal{Q})$ be a stratified pair of programs. Note $(\mathcal{P}_3, \mathcal{Q}_3)$ is also a stratified pair and that $(\mathcal{P} \cup \mathcal{Q})_3 = \mathcal{P}_3 \cup \mathcal{Q}_3$.

Prove by transfinite induction on ordinals η that if r is a relation symbol appearing in a rule in $\mathcal{P}$ and t is a tuple of terms, then

$$r(t) \in \mathbf{T}_{\mathcal{P}_3}\!\uparrow\!\eta \quad \text{if and only if} \quad r(t) \in \mathbf{T}_{\mathcal{P}_3 \cup \mathcal{Q}_3}\!\uparrow\!\eta \quad \text{and}$$
$$r_?(t) \in \mathbf{T}_{\mathcal{P}_3}\!\uparrow\!\eta \quad \text{if and only if} \quad r_?(t) \in \mathbf{T}_{\mathcal{P}_3 \cup \mathcal{Q}_3}\!\uparrow\!\eta$$

There are three cases:

- $\eta = 0$: By definition, $\mathbf{T}_{\mathcal{P}_3}\!\uparrow\!0 = \perp_L$, and $\mathbf{T}_{\mathcal{P}_3 \cup \mathcal{Q}_3}\!\uparrow\!0 = \perp_{L'}$, where L is the lattice of sets of ground atoms of $\mathcal{P}_3$, and L' is the lattice of sets of ground atoms of $\mathcal{P}_3 \cup \mathcal{Q}_3$. However, since r appears in $\mathcal{P}$, for each term t, $r_?(t) \in \perp_L$, $r_?(t) \in \perp_{L'}$, $r(t) \notin \perp_L$, and $r(t) \notin \perp_{L'}$.
- η *is a successor ordinal* $\mu + 1$: We shall show that if $r(t) \in \mathbf{T}_{\mathcal{P}_3 \cup \mathcal{Q}_3}\!\uparrow\!\eta$ then $r(t) \in \mathbf{T}_{\mathcal{P}_3}\!\uparrow\!\eta$. The other three cases are similar

or easier. As inductive hypothesis, assume the result for μ and for all relations appearing in $\mathcal{P}$.

If $r(t) \in \mathbf{T}_{\mathcal{P}_3 \cup \mathcal{Q}_3}$, then there is a ground instance of a rule of $\mathcal{P}_3 \cup \mathcal{Q}_3$

$$\rho \;=\; r(t) \leftarrow b_1, \ldots, b_k, \neg c_{1?}, \ldots, \neg c_{m?}$$

where each $b_i \in \mathbf{T}_{\mathcal{P}_3 \cup \mathcal{Q}_3} \uparrow \mu$ and each $c_{j?} \notin \mathbf{T}_{\mathcal{P}_3 \cup \mathcal{Q}_3} \uparrow \mu$.

Since $(\mathcal{P}_3, \mathcal{Q}_3)$ is a stratified pair, ρ is a (ground instance of a) rule of $\mathcal{P}_3$. By inductive hypothesis, each $b_i \in \mathbf{T}_{\mathcal{P}_3} \uparrow \mu$ and each $c_{j?} \notin \mathbf{T}_{\mathcal{P}_3} \uparrow \mu$. Hence $r(t) \in \mathbf{T}_{\mathcal{P}_3} \uparrow \eta$.

- η a limit ordinal: By inductive hypothesis, the result is true for all $\mu < \eta$. It follows that the result is true for $\mathbf{T}_{\mathcal{P}_3} \uparrow \eta = \sqcup \{\mathbf{T}_{\mathcal{P}_3} \uparrow \mu : \mu < \eta\}$ and $\mathbf{T}_{\mathcal{P}_3 \cup \mathcal{Q}_3} \uparrow \eta = \sqcup \{\mathbf{T}_{\mathcal{P}_3 \cup \mathcal{Q}_3} \uparrow \mu : \mu < \eta\}$ since those are formed by taking intersections and unions.

(2) Suppose $(\mathcal{P}, \mathcal{Q})$ is a stratified pair of programs, $\mathcal{P}$ and $\mathcal{P}'$ involve exactly the same symbols, and $\mathbf{F}_\mathcal{P} = \mathbf{F}_{\mathcal{P}'}$. So $(\mathcal{P}', \mathcal{Q})$ is also a stratified pair. Observe also that $(\mathcal{P} \cup \mathcal{Q})_3 = \mathcal{P}_3 \cup \mathcal{Q}_3$ and $(\mathcal{P}' \cup \mathcal{Q})_3 = \mathcal{P}'_3 \cup \mathcal{Q}_3$. Notice that, because of the stratification, for any set of ground atoms I of $\mathcal{P}_3$ and ground atoms J of $\mathcal{Q}_3$ (where the underlying terms may come from either $\mathcal{P}$ or $\mathcal{Q}$),

$$\mathbf{T}_{\mathcal{P}_3 \cup \mathcal{Q}_3}(I \cup J) = \mathbf{T}_{\mathcal{P}_3}(I) \cup \mathbf{T}_{\mathcal{Q}_3}(I \cup J).$$

Furthermore, if $(I \cup J)$ is a fixed point of $\mathbf{T}_{\mathcal{P}_3 \cup \mathcal{Q}_3}$, then

$$I = \mathbf{T}_{\mathcal{P}_3}(I) \quad \text{and} \quad J = \mathbf{T}_{\mathcal{Q}_3}(I \cup J).$$

Let $I = \{r(t), r_?(t) \in \mathbf{F}_{\mathcal{P} \cup \mathcal{Q}} : r \text{ appears in } \mathcal{P}\}$ and let $J = \mathbf{F}_{\mathcal{P} \cup \mathcal{Q}} - I$. By part (1), $I = \mathbf{F}_\mathcal{P}$, and, by assumption, $\mathbf{F}_{\mathcal{P}'} = \mathbf{F}_\mathcal{P} = I$.

$$\begin{aligned}
\mathbf{T}_{\mathcal{P}'_3 \cup \mathcal{Q}_3}(\mathbf{F}_{\mathcal{P} \cup \mathcal{Q}}) &= \mathbf{T}_{\mathcal{P}'_3}(I) \cup \mathbf{T}_{\mathcal{Q}_3}(I \cup J) \\
&= \mathbf{T}_{\mathcal{P}'_3}(\mathbf{F}_{\mathcal{P}'}) \cup \mathbf{T}_{\mathcal{Q}_3}(I \cup J) \\
&= \mathbf{F}_{\mathcal{P}'} \cup \mathbf{T}_{\mathcal{Q}_3}(I \cup J) \\
&= \mathbf{F}_\mathcal{P} \cup \mathbf{T}_{\mathcal{Q}_3}(I \cup J) \\
&= \mathbf{F}_\mathcal{P} \cup \mathbf{T}_{\mathcal{Q}_3}(\mathbf{F}_{\mathcal{P} \cup \mathcal{Q}}) \\
&= \mathbf{F}_\mathcal{P} \cup J \\
&= I \cup J \\
&= \mathbf{F}_{\mathcal{P} \cup \mathcal{Q}}.
\end{aligned}$$

Since $\mathbf{T}_{\mathcal{P}'_3 \cup \mathcal{Q}}(\mathbf{F}_{\mathcal{P} \cup \mathcal{Q}}) = \mathbf{F}_{\mathcal{P} \cup \mathcal{Q}}$, and since $\mathbf{F}_{\mathcal{P}' \cup \mathcal{Q}}$ is the $\sqsubseteq$-least fixed point of $\mathbf{T}_{\mathcal{P}'_3 \cup \mathcal{Q}}$, $\mathbf{F}_{\mathcal{P}' \cup \mathcal{Q}} \sqsubseteq \mathbf{F}_{\mathcal{P} \cup \mathcal{Q}}$. Similarly, $\mathbf{F}_{\mathcal{P} \cup \mathcal{Q}} \sqsubseteq \mathbf{F}_{\mathcal{P}' \cup \mathcal{Q}}$. So $\mathbf{F}_{\mathcal{P} \cup \mathcal{Q}} = \mathbf{F}_{\mathcal{P}' \cup \mathcal{Q}}$. ∎

Exercise 19. Show that the well-founded semantics satisfies the principle of stratification. (**Suggestion:** For part (2), use the characterization of $\mathbf{WF}_\mathcal{P}$ using unfounded sets. Show that, for $\mathcal{P}, \mathcal{P}'$, and $\mathcal{Q}$ as in the definition of the principle and U a set of ground atoms $r_?(t)$, if U is unfounded over $\mathbf{WF}_{\mathcal{P} \cup \mathcal{Q}}$ with respect to program $\mathcal{P}' \cup \mathcal{Q}$, then $U \cap \mathbf{WF}_{\mathcal{P} \cup \mathcal{Q}} = \emptyset$. To do this, break U into two parts, $U_\mathcal{P}$, the set of literals $r_?(t) \in U$ where r occurs in $\mathcal{P}$, and $U_\mathcal{Q} = U - U_\mathcal{P}$.)

Exercise 20. Show that the van Emden Kowalski semantics (for definite programs) and the perfect model semantics (for stratified programs) satisfy the principle of stratification.

Exercise 21
(1) Does classical logic satisfy the principle of stratification?
(2) Does the minimal model semantics satisfy part (1) of the principle of stratification? Does it satisfy part (2)?

Exercise 22. Develop one of the semantics suggested in Subsection 6.8. Compare it to all the semantics discussed in this chapter and evaluate it based upon all the criteria discussed in this section.

9 Some comments on expressive powers

The expressive powers of the standard logic programming semantics have been fairly actively studied. The study uses techniques of complexity theory and generalized recursion theory rather heavily. In fact, the study of expressive powers of these languages is mostly the study of how difficult the set of inferences is to compute, with the more expressive semantics being more difficult to compute. We shall give just some introductory results here.

Recall that we defined a set R of ground terms to be definable in a semantics $\mathcal{S}$ if, for some finite logic program $\mathcal{P}$ and some relation symbol r of $\mathcal{P}$,

$$R = \{t : r(t) \in Inferences_\mathcal{S}(\mathcal{P})\}.$$

There are other definitions of definability also used in the literature (e.g., uniform definability over finite databases), but we limit attention here to this one notion. Let $\mathcal{P}$ be any finite logic program. The set of all ground terms of $\mathcal{P}$ can be effectively enumerated, for example, by Gödel numbering.

Theorem 9.1 *Suppose a set R of terms is definable in the minimal model semantics. Then R is r.e.*

Proof Fix a program $\mathcal{P}$ where

$$R = \{t : r(t) \in Inferences_{minimal\ model}(\mathcal{P})\}.$$

Show that $r(t)$ holds in every minimal Herbrand model of $\mathcal{P}$ if and only if $r(t)$ holds in every Herbrand model of $\mathcal{P}$. For if there is a model $\mathcal{M}$ of $\mathcal{P} \cup \{\neg r(t)\}$, then, by Theorem 6.4, there is a minimal model $\mathcal{N}$ of $\mathcal{P}$, $\mathcal{N} \subseteq \mathcal{M}$, and obviously, $r(t) \notin \mathcal{N}$.

Let $\mathcal{P}^g$ be the ground instantiation of $\mathcal{P}$. The atom $r(t)$ holds in every model of $\mathcal{P}$ if and only if $r(t)$ holds in every model of $\mathcal{P}^g$. By the completeness theorem, $r(t)$ holds in every model of $\mathcal{P}^g$ if and only if there is a (finite) proof of $r(t)$ from $\mathcal{P}^g$. But the set of all $r(t)$ for which such a finite proof exists is r.e. ∎

The converse to Lemma 9.1 is also true. Smullyan (1956a, 1956b) (using a different formalism), and independently Andreka and Nemeti (1978), showed that a set of ground terms is r.e. if and only if it is definable in the van Emden Kowalski semantics, and thus if and only if it is definable by a definite program in the minimal model semantics. (To prove completeness, one can write a definite program, using function symbols, where the program "simulates" the moves of a Turing machine.)

Exercise 23. How complicated is the set of negative literals inferred from a logic program in the minimal model semantics? Answer this separately (1) for definite programs, and (2) for normal programs.

Apt and Blair (1990) showed further that a set of ground terms is definable by a stratified program in the perfect model semantics if and only if it is arithmetic, that is, definable over the natural numbers by a first order formula (with functions $+$ and $\cdot$).

The two program completion semantics, the well-founded semantics, and the stable semantics are more expressive.

Proposition 9.2 *If a set S is definable in the 3-valued program completion semantics, it is definable in the 2-valued program completion semantics. If it is definable in the well-founded semantics, it is definable in the stable semantics.*

Proof Suppose $S = \{t : r(t) \in \text{Inferences}_{\text{3-valued program completion}}(\mathcal{P})\}$. Note that, in general,

$$S \neq \{t : r(t) \in \text{Inferences}_{\text{2-valued program completion}}(\mathcal{P})\}.$$

But, by definition of the 3-valued program completion semantics,

$$S = \{t : r(t) \in \text{Inferences}_{\text{2-valued program completion}}(\mathcal{P}_3)\},$$

so S is definable in the 2-valued program completion semantics.

The proof for the well-founded semantics and the stable semantics is analogous. ∎

Definition 9.3 *Let $\mathcal{P}$ be a normal logic program. The double negative transformation of $\mathcal{P}$ is the program $\mathcal{P}^{\neg\neg}$ formed from $\mathcal{P}$ as follows. To* simplify the description, we treat proposition letters here as 0-ary relation symbols.

- For each relation symbol r of $\mathcal{P}$, add a new relation symbol r^-, and include in $\mathcal{P}^{\neg\neg}$ the rule $r^-(\vec{X}) \leftarrow \neg r(\vec{X})$, where $\vec{X}$ is a tuple of variables of the appropriate length for r.
- For each rule
$$r(\vec{t}) \leftarrow s_1(\vec{t_1}), \ldots, s_k(\vec{t_k}), \neg s'_1(\vec{t_1}'), \ldots, \neg s'_l(\vec{t_l}')$$
of $\mathcal{P}$, include in $\mathcal{P}^{\neg\neg}$ the rule
$$r(\vec{t}) \leftarrow \neg s_1^-(\vec{t_1}), \ldots, \neg s_k^-(\vec{t_k}), \neg s'_1(\vec{t_1}'), \ldots, \neg s'_l(\vec{t_l}').$$

Lemma 9.4 *Let $\mathcal{P}$ be a normal logic program. Then*

(1) $\text{Inferences}_{\text{2-valued program completion}}(\mathcal{P})$
$= \text{Inferences}_{\text{2-valued program completion}}(\mathcal{P}^{\neg\neg}),$
and
$\text{Inferences}_{\text{3-valued program completion}}(\mathcal{P})$
$= \text{Inferences}_{\text{3-valued program completion}}(\mathcal{P}^{\neg\neg}).$

(2) $\text{Inferences}_{\text{2-valued program completion}}(\mathcal{P}^{\neg\neg}) = \text{Inferences}_{\text{stable}}(\mathcal{P}^{\neg\neg}),$
and
$\text{Inferences}_{\text{3-valued program completion}}(\mathcal{P}^{\neg\neg})$
$= \text{Inferences}_{\text{well-founded}}(\mathcal{P}^{\neg\neg}).$

Proof (1) First consider the two 2-valued semantics. For any Herbrand interpretation $\mathcal{M}$ for $\mathcal{P}$, let $\mathcal{M}^+$ be the Herbrand interpretation for $\mathcal{P}^{\neg\neg}$ given by
$$\mathcal{M}^+ = \mathcal{M} \cup \{r^-(\vec{t}) : r(\vec{t}) \notin \mathcal{M}\},$$
where r above ranges over the relation symbols of $\mathcal{P}$ and, if r is k-ary, $\vec{t}$ ranges over all k-tuples of ground terms of $\mathcal{P}$. For $\mathcal{N}$ any Herbrand interpretation for $\mathcal{P}^{\neg\neg}$, let $\mathcal{N}^-$ be the Herbrand interpretation for $\mathcal{P}$ given by
$$\mathcal{N}^- = \{s(\vec{t}) \in \mathcal{N} : s \text{ is a predicate of } \mathcal{P}\}.$$
(Thus $\mathcal{N}^-$ is the *reduct* of $\mathcal{N}$ to the language of $\mathcal{P}$.)
Now show that if $\mathcal{M}$ is a fixed point of $\mathbf{T}_{\mathcal{P}}$, then $\mathcal{M}^+$ is a fixed point of $\mathbf{T}_{\mathcal{P}^{\neg\neg}}$, and if $\mathcal{N}$ is a fixed point of $\mathbf{T}_{\mathcal{P}^{\neg\neg}}$, then $\mathcal{N}^-$ is a fixed point of $\mathbf{T}_{\mathcal{P}}$. The desired result will follow immediately.
We show that if $\mathcal{N}$ is a fixed point of $\mathbf{T}_{\mathcal{P}^{\neg\neg}}$, then $\mathcal{N}^-$ is a fixed point of $\mathbf{T}_{\mathcal{P}}$; the other half we leave for the reader. So suppose $\mathcal{N}$ is a fixed point. For each relation symbol $r(t)$, there is only one rule in $\mathcal{P}^{\neg\neg}$ defining r^-: the rule $r^-(\vec{X}) \leftarrow \neg r(\vec{X})$. Thus for any relation symbol r – say r is k-ary – and any k-tuple $\vec{t}$ of ground terms,
$$r^-(\vec{t}) \in \mathcal{N} \text{ if and only if } r(\vec{t}) \notin \mathcal{N}.$$

Thus the body of a rule

$$r(\vec{t}) \leftarrow s_1(\vec{t_1}), \ldots, s_k(\vec{t_k}), \neg s'_1(\vec{t_1}'), \ldots, \neg s'_l(\vec{t_l}')$$

is true in $\mathcal{N}^-$ if and only if the body of the same rule

$$r(\vec{t}) \leftarrow s_1(\vec{t_1}), \ldots, s_k(\vec{t_k}), \neg s'_1(\vec{t_1}'), \ldots, \neg s'_l(\vec{t_l}')$$

is true in $\mathcal{N}$, if and only if the body of the corresponding rule

$$r(\vec{t}) \leftarrow \neg s_1^-(\vec{t_1}), \ldots, \neg s_k^-(\vec{t_k}), \neg s'_1(\vec{t_1}'), \ldots, \neg s'_l(\vec{t_l}').$$

is true in $\mathcal{N}$. It follows immediately that $\mathcal{N}^-$ is a fixed point of $\mathbf{T}_{\mathcal{P}}$. For the 3-valued program completion semantics, note that $(\mathcal{P}^{\neg\neg})_3 = (\mathcal{P}_3)^{\neg\neg}$. The result for program $\mathcal{P}$ in the 3-valued program completion then follows from the result for $\mathcal{P}_3$ in the 2-valued program completion semantics.

(2) Since no rule $\mathcal{P}^{\neg\neg}$ has any positive subgoals, by Proposition 6.38, the stable semantics makes the same inferences from $\mathcal{P}^{\neg\neg}$ as does the 2-valued program completion semantics, and the well-founded semantics makes the same inferences from $\mathcal{P}^{\neg\neg}$ as does the 3-valued program completion semantics. ∎

It follows that

Theorem 9.5 *If a set S is definable in the 3-valued program completion semantics, it is definable in the well-founded semantics. If it is definable in the 2-valued program completion semantics, it is definable in the stable semantics.*

In fact, the converses to Proposition 9.2 and Theorem 9.5 also hold. Exactly the same sets of terms are definable in all four semantics: the so-called Π_1^1 sets of natural numbers (see (Fitting 1985; Schlipf 1990; Marek et al. 1992)). Of course, there may be a fairly simple program which defines a set R of terms in the stable semantics, while any program which defines R in the 3-valued program completion semantics is substantially more complicated and less intuitive. But Proposition 9.2 and Theorem 9.5 also hold in much more general contexts than we have presented here, and for some of these contexts the converses fail.

Bibliography

1. H. Andreka and I. Nemeti (1978). The generalized completeness of Horn predicate logic as a programming language. *Acta Cybernetica*, *4(1)*, 3–10.
2. K. R. Apt and H. Blair (1990). Arithmetic classification of perfect models of stratified programs. *Fund. Informaticae*, *13(1)*, 1–71.
3. K. R. Apt, H. Blair, and A. Walker (1988). Towards a theory of declarative knowledge. In J. Minker, editor, *Foundations of Deductive Databases and Logic Programming*, Morgan Kaufmann, Los Altos, CA, 89–148.

4. C. Baral, J. Lobo, and J. Minker (1991). WF^3: a semantics for negation in normal disjunctive logic programs. In Z. W. Ras and M. Zemankova, editors, *Methodologies for Intelligent Systems: Proceedings of the 6th International Symposium*, Springer-Verlag, Berlin, 459–468.
5. A. Bondarenko, F. Toni, and R. A. Kowalski (1993). An assumption-based framework for non-monotonic reasoning. In L. Pereira and A. Nerode, *Logic Programming and Non-monotonic Reasoning, Proc. of the 2nd. International Workshop*, 171–189.
6. M. Cadoli and M. Schaerf (1993). A survey of complexity results for non-monotonic logics. *Journal of Logic Programming, 17(2,3,4)*, 127–160.
7. A. Chandra and D. Harel (1982). Structure and complexity of relational queries. *J. Computer and Systems Sciences, 25(1)*, 99–128.
8. A. Chandra and D. Harel (1985). Horn clause queries and generalizations. *JACM, 29(3)*, 841–862.
9. J. Chen and S. Kundu (1991). The strong semantics for logic programs. In Z. W. Ras and M. Zemankova, editors, *Methodologies for Intelligent Systems: Proceedings of the 6th International Symposium*, Springer-Verlag, Berlin, 490–499.
10. K. L. Clark (1978). Negation as failure. In Gallaire and Minker, editors, *Logic and Databases*, Plenum Press, New York, 293–322.
11. M. Davis (1980). The mathematics of non-monotonic reasoning. *Artificial Intelligence, 28*, 73–80.
12. J. Dix (1990). Cumulativity and rationality in semantics of normal logic programs. In J. Dix, K. P. Jantke, and P. H. Schmitt, editors, *Nonmonotonic and Inductive Logic: Proceedings of the 1st International Workshop*, Springer-Verlag, Berlin, 13–37.
13. P. M. Dung (1991). Negation as hypothesis: an abductive foundation for logic programming. *Proc. 8th International Conference on Logic Programming*, MIT Press, Cambridge, MA.
14. P. M. Dung. (1992). On the relations between stable and well-founded semantics. *Theoretical Computer Science*, 105: 7–25.
15. K. Eshghi and R. A. Kowalski (1989). Abduction compared with negation as failure. *Proc. 6th International Conference on Logic Programming*, MIT Press, Cambridge, MA, 234–254.
16. M. Fitting (1985). A Kripke-Kleene semantics for logic programs. *Journal of Logic Programming, 2(4)*, 295–312.
17. M. Fitting (1993). The family of stable models. *Journal of Logic Programming, 17(2,3,4)*, 197–226.
18. M. Gelfond (1987). On stratified autoepistemic theories. In *Proc. AAAI*.
19. M. Gelfond and V. Lifschitz (1988). The stable model semantics for

logic programming. In *Proc. 5th Int'l Conf. Symp. on Logic Programming*, vol. 2, 1070–1080.

20. A. C. Kakas and P. Mancarella (1990). Abductive logic programming. In *Proceedings of the Workshop: Logic Programming and Non-Monotonic Logic* (conducted with the North American Conference on Logic Programming), 49–61.

21. A. C. Kakas and P. Mancarella (1991). Negation as stable hypotheses. In A. Nerode, W. Marek, and V.S. Subramahnian, editors, *Logic Programming and Non-monotonic reasoning: Proceedings of the First International Workshop*, MIT Press, Cambridge, MA, 275–288.

22. A. C. Kakas and P. Mancarella (1991). Stable theories for logic programs. In V.J. Saraswat and K. Ueda, editors, *Proc. 1991 International Symposium on Logic Programming*, MIT Press, Cambridge, MA, 85–100.

23. P. Kolaitis (1991). The expressive power of stratified programs. *Inf. Comput.*, *90 (1)*, 50–66.

24. P. Kolaitis and C. Papadimitriou (1991). Why not Negation by Fixpoint? *JCSS*, *43*, 125–144.

25. S. Kraus, D. Lehman, and M. Magidor (1990). Nonmonotonic reasoning, preferential models, and cumulative logics. *Artificial Intelligence 44*, 167–207.

26. K. Kunen (1987). Negation in logic programming. *Journal of Logic Programming*, *4(4)*, 289–308.

27. V. Lifschitz (1988). On the declarative semantics of logic programs with negation. In J. Minker, editor, *Foundations of Deductive Databases and Logic Programming*, Morgan Kaufmann, Los Altos, CA, 177–192.

28. J. W. Lloyd (1987). *Foundations of Logic Programming* (2nd edn). Springer-Verlag, New York.

29. J. McCarthy (1980). Circumscription – a form of non-monotonic reasoning. *Artificial Intelligence*, *13*, 27–39.

30. W. Marek and M. Truszczyński (1991). Autoepistemic logic. *Journal of the ACM*, *38(3)*, 588–619.

31. W. Marek, A. Nerode, and J. Remmel (1992). How complicated is the set of stable models of a recursive logic program? *Annals of Pure and Applied Logic*, *56*, 119–135.

32. J. Minker (1982). On indefinite databases and the closed world assumption. In *Sixth Conference on Automated Deduction*, Springer-Verlag, New York, 292–308.

33. T. C. Przymusiński (1988). On the declarative semantics of deductive databases and logic programs. In J. Minker, editor, *Foundations of Deductive Databases and Logic Programming*, Morgan Kaufmann, Los Altos, CA, 193–216.

34. T. C. Przymusiński (1989). Every logic program has a natural stratification and an iterated fixed point model. In *Eighth ACM Symposium on Principles of Database Systems*, 11–21.
35. K. A. Ross (1989). A procedural semantics for well-founded negation in logic programs. In *Eighth ACM Symposium on Principles of Database Systems*, 22–33.
36. D. Saccà and C. Zaniolo (1990). Stable models and non-determinism in logic programs with negation. In *Ninth ACM Symposium on Principles of Database Systems*, 205–217.
37. J. S. Schlipf (1990). The expressive powers of the logic programming semantics. *Ninth ACM Symposium on Principles of Database Systems*, 196–204, 1990.
38. J. Schlipf (1991). Representing epistemic intervals in logic programming. In A. Nerode, W. Marek, and V.S. Subramahnian, editors, *Logic Programming and Non-monotonic reasoning: Proceedings of the First International Workshop*, MIT Press, Cambridge, Massachusetts, 133–147.
39. J. Schlipf (1992). Formalizing a logic for logic programming. *Annals of Mathematics and Artificial Intelligence*, 5, 279–302.
40. J. C. Shepherdson (1988). Negation in logic programming. In J. Minker, editor, *Foundations of Deductive Databases and Logic Programming*, Morgan Kaufmann, Los Altos, CA, 19–88.
41. R. M. Smullyan (1956a). Elementary formal system (abstract). *Bull. AMS*, *62*, 600.
42. R. M. Smullyan (1956b). On definability by recursion (abstract). *Bull. AMS*, *62*, 601.
43. M. H. Van Emden and R. A. Kowalski (1976). The semantics of predicate logic as a programming language. *JACM*, *23(4)*, 733–742.
44. A. Van Gelder (1989a). Negation as failure using tight derivations for general logic programs. *Journal of Logic Programming* 6, 109–133.
45. A. Van Gelder (1989b). The alternating fixpoint of logic programs with negation. In *Eighth ACM Symposium on Principles of Database Systems*, 1–10.
46. A. Van Gelder and J. S. Schlipf (1993). Common sense axiomatizations for logic programs. *Journal of Logic Programming*, *17(2,3,4)*, 161–196.
47. A. Van Gelder, K. A. Ross, and J. S. Schlipf (1991). The well-founded semantics for general logic programs. *Journal of the ACM*, *38(3)*, pages 620–650.

2
From Concurrent Logic Programming to Concurrent Constraint Programming

Frank S. de Boer
Free University Amsterdam

Catuscia Palamidessi
Università di Genova

Abstract

The endeavor to extend logic programming to a language suitable for concurrent systems has stimulated in the last decade an intensive research, resulting in a large variety of proposals. A common feature of the various approaches is the attempt to define mechanisms for concurrency within the logical paradigm, the driving ideal being the balance between expressiveness and declarative reading. In this survey we present the motivations, the principal lines along which the field has developed, the various paradigms which have been proposed, and the main approaches to the semantic foundations.

1 Introduction

Among the various reasons which have contributed to the popularity of logic programming, one is the opinion that it is an *inherently parallel* language, therefore suitable for parallel and distributed architectures. The pure language can already be regarded as a model for parallel computation: in the so-called *process interpretation* (van Emden and de Lucena 1982; Shapiro 1983), the goal is seen as a system of parallel processes. The single process evolves via resolution steps; different clausal definitions for the same predicate provide nondeterminism and the shared variables among processes represent communication channels.

However, in a concurrent system processes *interact* with each other[8], hence a concurrent language should embody some synchronization mechanisms and constructs to specify the dependency of choices upon the environment. Furthermore once a certain choice is selected, the process cannot backtrack. This principle is in contrast with the kind of nondeterminism of

---

[8]The dichotomy sequential programming/concurrent programming is closely related to the distinction between transformational systems and interactive (or reactive) systems, see (Harel and Pnueli 1985).

logic programming, according to which all possible solutions are pursued (*don't know nondeterminism*).

For more than a decade there has been an active line of research aimed at extending logic programming with mechanisms for concurrency.

1.1 Don't know versus don't care nondeterminism

In concurrent systems a process cannot "undo" a choice, even if it proves to be wrong (for instance because the process gets stuck). The reason is that a choice, and the actions done afterwards, have already influenced the environment: to maintain consistency the whole system should backtrack, but this is too inefficient. It is rather preferable to provide the language with mechanisms to control the choices, so to avoid as far as possible that wrong decisions are taken. The most common such mechanism is the *guard*: a condition associated with each alternative which is tested before selecting that alternative.

Almost all the concurrent logic languages and concurrent constraint languages adopt *don't care nondeterminism*: clauses are provided with a guard part and a process chooses (*commits* to) only one clause, provided its *guard part* is satisfied.

The few languages which maintain don't know nondeterminism, thus preserving the characteristics of logic programming to compute all solutions, are Distributed Logic (Monteiro 1981,1982), the language of Generalized Horn Clauses (Falaschi *et al.* 1984), Delta PROLOG (Pereira and Nasr 1984) and the language $cc(\downarrow, \rightarrow, \Rightarrow)$ (Saraswat 1986, 1989). Andorra PROLOG (Haridi and Brand 1988; Haridi 1990; Haridi and Janson 1990, 1991; Haridi *et al.* 1992) combines don't care and don't know nondeterminism.

1.2 Synchronization and communication mechanisms

Roughly, a synchronization mechanism specifies when a process can be activated. In many concurrent paradigms synchronization is in symbiosis with communication. This is the case also in most concurrent logic languages. Some of the first concurrent logic languages, the Relational Language (Clark and Gregory 1981), Concurrent PROLOG (Shapiro 1983, 1986, 1988), the language of Guarded Horn Clauses (Ueda 1987, 1988) and PARLOG (Clark and Gregory 1986; Gregory 1987), enforced synchronization by introducing a directionality on the communication channels. In this way, some processes are seen as *producers* (of bindings) on a certain variable, others as *consumers*. The advantage of such a solution is that the resulting language is quite "faithful" to the (parallel) execution model of logic programming: only the unification mechanism needs to be modified.

The kind of communication mechanism described above is asynchronous[9], in the sense that the process producing a binding and the process consuming (or, more precisely, testing) the binding, perform their actions at different times. As opposite to this principle, synchronous communication requires the partners to exchange the information *simultaneously*. This latter mechanism has been adopted in Distributed Logic, Generalized Horn Clauses, Delta PROLOG and the language of Communicating Clauses (Jacquet and Monteiro 1990, 1991, 1992).

A rather different approach to synchronization has been proposed in P-PROLOG (Yang 1986; Yang and Aiso 1986). The idea is to suspend a process until enough bindings are produced so that the number of consistent alternatives for the next step reduces to one. This solution is semantically clean, in the sense that it does not modify the declarative reading of the language. Of course, the disadvantage is that only deterministic processes can be specified in this way. Such a mechanism is also embodied in Andorra PROLOG and in the ALPS languages (Maher 1987).

1.3 The influence of Constraint Logic Programming

The ALPS paradigm was introduced as the concurrent version of Constraint Logic Programming (Jaffar and Lassez 1987). The main novelty was the *validation* case for the commitment rule. In short, this rule allows a process to make a derivation step using a certain clause only if the constraint established till that moment *entails* the constraint associated with the guard of the clause. Essentially this is the "natural evolution" of the GHC commitment rule, but the way it is formulated is much more elegant and combines more naturally with the underlying logic paradigm.

This intuition had a strong impact on the field: it was the precursor of the ask-rule, which is one of the central ideas of concurrent constraint programming and which has become the basic synchronization mechanism of concurrent logic/constraint languages.

1.4 Concurrent constraint programming

Concurrent constraint programming (ccp) (Saraswat 1989; Saraswat and Rinard 1990; Saraswat *et al.* 1991) presents a new perspective on the underlying philosophy of logic programming. In fact it is based on extralogical *operators* typical of the concurrent paradigms, like CCS (Milner 1980), TCSP (Brookes *et al.* 1984) and ACP (Bergstra and Klop 1984, 1986); in particular, the *choice* $(+)$, the *action prefixing* $(\rightarrow)$, and the *hiding* operator $(\exists)$. Additionally, concurrent constraint programming embodies explicit mechanisms for communication and synchronization consisting of two kinds of actions, *ask* and *tell*. Also in the concurrent logic languages

---

[9]The terminology *synchronous/asynchronous communication* must not be confused with *synchronization* in the sense of suspending/resuming processes. A synchronization mechanism can be based on asynchronous communication.

these control features were present, but they were "hidden" in various ways, as explained in Section 1.

There are many advantages in an explicit representation of these concurrency control mechanisms by means of operators. First of all, it is the basis for the definition of a calculus and for an algebraic treatment of processes. Furthermore, the explicit representation allows us to separate each mechanism from the others, thus permitting a better understanding of the laws of its behavior. For instance, one of the main problems in studying the semantics of concurrent logic programming is that the choice mechanism is combined with recursion, since a clause is in general a recursive definition. Another advantage is that the standard tools developed in the theory of concurrency can be applied more easily. Finally, a "reconciliation" with the declarative principles of logic programming is made more feasible. For instance, the renaming mechanism and the behavior of the *ask* and *tell* actions can be described in a logical way.

Concurrent constraint programming has become rapidly very popular and is regarded as the reference formalism in this area.

1.5 Semantics

Compared to the pure logic and constraint languages, the introduction of the various synchronization and control mechanisms induces a more complicated semantics. At the operational level, these constructs can be appropriately described by means of a *transition system* in the style of SOS (Plotkin 1981), see for instance (Saraswat 1987a, 1987b).

However, problems arise when developing semantics which account for the declarative nature. There are various differences with respect to the pure languages, which make problematic a simple extension of the standard model-theoretic approach. First of all, due to the presence of the synchronization mechanisms, not all the computation steps are allowed and therefore the *success set* becomes smaller. Levi and Palamidessi (1985, 1987) developed a model-theoretic semantics for characterizing the success set of concurrent logic languages. The basic idea was to annotate the data structures in order to model the producer/consumer discipline, thus obtaining a sort of input output semantics. However not all kinds of concurrent computations could be modeled in this way.

Another problem arises from the don't care nondeterminism: the concept of failure changes from *universal* to *existential*. Furthermore, a new kind of termination must be taken into account: *deadlock*, namely the situation in which all processes are *suspended*, waiting for the conditions for synchronization to become validated. Note that don't care nondeterminism means absence of backtracking, hence non-successful computations are particularly relevant in interactive systems: a system which can deadlock/fail cannot be considered equivalent to a system deadlock/failure free, even if the set of successful computations is the same.

The characterization of non-successful computations requires even more complicated semantic structures and operators, which are difficult to bring into accordance with the logical flavor. However, if it is problematic to maintain the logical flavor, yet we can preserve the other important aspect of the model-theoretic semantics: the compositionality. This amounts to shifting to a denotational approach.

The first proposals of denotational semantics for concurrent logic languages were inspired by the classical concepts of concurrency theory. In particular, concerning the choice of the semantical domains: synchronization trees and failure sets. For example, De Bakker and Kok (1988,1990) and de Boer *et al.* (1989,1991) used tree-like structures labeled with functions; Gabbrielli and Levi (1990) considered trees labeled with constraints; Saraswat and Rinard (1990) used similar structures modulo equivalence relations based on bisimulation; Gerth *et al.* (1988) and Gaifman *et al.* (1989) used refusal pairs as in the failure semantics (Brookes *et al.* 1984), although in (Gerth *et al.* 1988) the refusal set is reduced to a singleton.

Tree-like structures, bisimulation and failure semantics encode, in various degrees of abstraction, the branching structure of a process. Such information is needed, in CCS, TCSP, ACP, and in languages with *synchronous* communication in general, to represent the dependency of a choice upon the environment, hence to model compositionally the choice and the parallel operators.

On the other hand, in concurrent logic programming and in concurrent constraint programming communication is *asynchronous*. Therefore those branching structures are actually not needed: which actions are enabled does not depend upon the current state of the environment, but only upon the store. In a transition system this can be made explicit by adding a *passive* rule that does not exist in the classical concurrent paradigms: an *arbitrary* assumption of a step made by the environment. This amounts to considering all the possible interactions between the given process and arbitrary environments, and it leads to a simple compositional semantics, consisting of sequences of constraints labeled by assume/tell modes (de Boer and Palamidessi 1990a, 1991a). In this framework the parallel combination corresponds to "unzip" sequences, so that the assumptions of a process match with the actions of the other, and vice versa. Sequences of this kind (*reactive sequences*) were already one component of the semantics in (Gerth *et al.* 1988) and in (Gaifman *et al.* 1989), the other component being the refusal information.

Independently, an intriguing approach was developed by Saraswat *et al.* (1991). The basic idea consists of denoting processes by Scott's closure operators, which have the nice property of being representable via the set of their fixpoints. The operators of the language can then be described as operations on those sets. In particular, parallelism can be modeled simply by intersection. This semantics abstracts from deadlock, however it should be possible to extend it so to model deadlock as well.

The approach of Saraswat *et al.* (1991) is suitable for languages without atomic unification, i.e. languages which do not support the consistency check as a guard of a choice. In this context, the approach of de Boer and Palamidessi (1991a) is equivalent. The semantics of Saraswat *et al.* is more elegant as it exploits the monotonic evolution of the computations; on the other hand the approach of de Boer and Palamidessi is more general because it can be applied also to languages with atomic unification. Furthermore it can be easily adapted to any extension of ccp, provided that we remain in the framework of asynchronous communication (see de Boer *et al.* 1994).

1.6 Current trends

The application of category theory has brought new insight to the semantics of ccp. Panangaden *et al.* (1992) used the framework of hyperdoctrines to model the entailment relation. Saraswat (1992) showed that constraint systems form a Cartesian-closed category, thus establishing the basis for the solution of recursive domain equations and higher-order constraint languages.

Saraswat and Lincoln (1992) have developed a variant of ccp based on linear logic. Linear logic is particularly suitable to deal with concrete aspects of the computation, such as the consumption of resources. Furthermore, it seems to be an appropriate basis for the integration of logic programming and object-oriented programming (Andreoli and Pareschi 1991).

De Boer and Palamidessi (1992) have made a first step towards the algebraic axiomatization of ccp. Process algebra is one of the most widespread methods in concurrency theory both for specification and for program verification, and it has been intensively investigated in the synchronous case. In the asynchronous case however a lot of work remains to be done: modeling various notions of observables and capturing by axioms the specificity of the communication mechanisms of the various languages.

Recently there has been a lot of interest concerning expressiveness questions. Shapiro (1991, 1992) and de Boer and Palamidessi (1990b, 1991b, 1994) have done some preliminary work aimed at the development of a general theory to compare the expressive power of concurrent languages in general, and cc languages in particular. Such studies are meant to provide a formal basis of the design principles and implementation techniques.

Another interesting line of research is the application of abstract interpretation techniques to the analysis of ccp. A compositional approach to this problem has been developed by Falaschi *et al.* (1993). The idea is to consider a denotational semantics which correctly approximates the observables, without being correct in the classical sense. This allows us to consider simpler semantic structures which make the analysis more efficient.

An efficient distributed implementation of ccp requires the development of non-interleaving models. Montanari and Rossi (1991) have developed a first proposal for a "true concurrent" model. Such a model is based on *graph grammars*: the inference rules of the underlying constraint system and the (user specified) agents are expressed as *graph productions*. A computation is a sequence of graph rewriting steps, and it is possible to associate with it an *occurrence net*, a kind of Petri net which expresses *causality information*.

1.7 Plan of the paper

For various reasons, which include implementability considerations and smooth integration with the logical paradigm, the asynchronous approach to communication, based on shared variables, has become more popular and more widespread in the design of concurrent logic languages. Therefore this paper will focus on the class of asynchronous languages.

The paper is organized as follows. Next section contains some preliminaries about logic programming. Section 3 presents the main lines along which the field of concurrent logic programming has developed; the motivations, the main ideas, the various proposals for a right balance between expressivity and faithfulness to the logical paradigm. In order to point out similarities and differences of the various languages, we try to be as uniform as possible in the formal description of their operational semantics, given via a transition system.

In Chapter 4 we review the innovative ideas introduced by the ALPS paradigm. Finally, in Section 5 the concurrent constraint paradigm is presented. For this class of languages we study in detail the denotational semantics. This will provide also a model for the concurrent logic languages, since concurrent constraint programming can be seen as a generalization of (most of) them. We review and compare the two main denotational approaches presented in the literature, one based on reactive sequences, the other on Scott's closure operators.

2 Preliminaries.

In this section we recall the terminology and the basic definitions of logic programming. For more details we refer to (Lloyd 1987; Apt 1990).

We assume given a finite set of *data constructors* $a, b, c, \ldots$ (constant symbols), $f, g, h, \ldots$ (function symbols), a finite set of *predicate symbols* $p, q, r, \ldots$ and a denumerable set Var of *variable symbols* $x, y, z, \ldots$. Terms $t, u, \ldots$ are constructed from data constructors and variables in the usual way. *Atoms* $A, B, \ldots$ are objects of the form $p(t_1, \ldots, t_n)$. Terms and atoms are called *ground* if they don't contain occurrences of variables. A *substitution* ϑ is a mapping from variables to terms such that the set $dom(\vartheta) = \{x \mid \vartheta(x) \neq x\}$ (*domain* of ϑ) is finite. The empty substitution (identity) is denoted by ϵ. The term $t\vartheta$ results from simultaneously replacing in t every variable x by $\vartheta(x)$. Thus substitutions can be lifted to functions from terms to terms. The composition $\vartheta\sigma$ of the substitutions ϑ and σ is defined as the functional composition. The pre-order $\leq$ on substitutions is defined by $\vartheta \leq \sigma$ iff there exists ϑ' such that $\vartheta\vartheta'' = \sigma$. We denote by $\simeq$ the associated equivalence relation. The application of a substitution to an atom is defined as $p(t_1,\ldots,t_n)\vartheta = p(t_1\vartheta,\ldots,t_n\vartheta)$. Given two atoms A and B, their most general unifier (unique modulo $\simeq$) $mgu(A,B)$ is the minimal substitution ϑ such that $A\vartheta = B\vartheta$.

A *definite clause* C is a formula $H \leftarrow B_1,\ldots,B_n$ $(n \geq 0)$, where H and the B_i's are atoms; H is the *head* of the clause and the sequence $B_1,\ldots,B_n$ is the *body*. If the body is empty then the clause is called a *unit clause*. The symbols "$\leftarrow$" and "," denote logical implication and logical conjunction respectively. Clauses are assumed to be universally closed. A *program* is a finite set of definite clauses $P = \{C_1,\ldots,C_n\}$. A *goal statement* G is a formula $\leftarrow A_1,\ldots,A_m$, where each A_i is an atom.

The standard operational semantics is defined as follows. A *derivation step* is a goal transformation

$$\leftarrow A_1,\ldots,A_i,\ldots,A_m \overset{\vartheta}{\longmapsto} \leftarrow (A_1,\ldots,A_{i_1},B_1,\ldots B_n,A_{i+1},\ldots,A_m)\vartheta$$

where $H \leftarrow B_1,\ldots B_n$ is a clause in P, such that $\vartheta = mgu(A_i,H)$. The clause must be renamed in order to avoid variable clashes with the goal (*renaming apart*). A *derivation* of $P \cup \{G\}$ is a possibly infinite sequence of goals $G_1,\ldots,G_n,\ldots$, where $G = G_1$, such that for all subsequent goals G_i, G_{i+1} we have $G_i \overset{\vartheta_i}{\longmapsto} G_{i+1}$ for some substitution ϑ_i. A derivation is *successful* if it is finite and it terminates with the empty goal ($\square$). The composition of all the associated substitutions is called *computed answer substitution*. A derivation *fails* if it is finite, maximal (i.e. it is not the prefix of a longer derivation) and it is not successful. A derivation is *fair* if it is either finite or every atom which occurs in some goal of the derivation is eventually selected for a derivation step.

The success set and finite failure set of a program P are defined[10] respectively by

$SS = \{A \mid A \text{ is ground and } P \cup \{\leftarrow A\} \text{ has a refutation}\}$,

$FF = \{A \mid A \text{ is ground and all fair derivations of } P \cup \{\leftarrow A\} \text{ fail}\}$.

3 Concurrent logic languages

In this section we discuss various approaches to turn logic programming into a language suitable for concurrent systems. First we consider the so-called *process interpretation* of the pure language, and then we show what mechanisms have to be added in order to model the basic concepts of concurrency.

---

[10]The failure set is usually defined in terms of a finitely failing SLD-tree; however, the definition that we consider here is equivalent to the standard one (Lassez and Maher, 1984).

C1	$\langle \leftarrow A; \vartheta \rangle \longrightarrow \langle \leftarrow \bar{B}; \vartheta\sigma \rangle$	*there exists $A' \leftarrow \bar{B} \in P_{ren}$ s.t. $\sigma = mgu(A\vartheta, A')$*
C2	$\dfrac{\langle \leftarrow \bar{A}; \vartheta \rangle \longrightarrow \langle \leftarrow \bar{A}'; \vartheta' \rangle}{\begin{array}{c}\langle \leftarrow \bar{A}, \bar{B}; \vartheta \rangle \longrightarrow \langle \leftarrow \bar{A}', \bar{B}; \vartheta' \rangle \\ \langle \leftarrow \bar{B}, \bar{A}; \vartheta \rangle \longrightarrow \langle \leftarrow \bar{B}, \bar{A}'; \vartheta' \rangle \end{array}}$	

Table 1 The transition system for logic programming

3.1 Process interpretation of logic programming

A derivation in logic programming can be viewed as the evolution of a dynamic network of parallel processes (van Emden and de Lucena 1982; Shapiro 1983). An atom in the goal corresponds to a process and a variable occurring in two or more atoms represents a communication channel between the corresponding processes. Thus a goal represents a network of parallel processes connected by communication channels. Communication is modeled by the instantiation of the variables during unification. The resolution step gives rise to a reconfiguration of the network by expanding a process (the selected atom) into a subnetwork of parallel processes (the atoms of the body of the selected clause), and establishing the links on the communication channels. Note that a variable can be instantiated to a term which contains other variables, hence also the connection structure can be modified.

We formalize the above described process view in terms of a *transition system*. In the following a sequence of objects (atoms, terms, variables etc.) $\chi_1, \ldots, \chi_n$ is denoted by $\bar{\chi}$.

Definition 3.1 (A transition system for logic programming) *The process semantics of logic programming is described by a transition system whose configurations are pairs $\langle G; \vartheta \rangle$, where G is a goal and ϑ is a substitution. The transition relation is defined by the rules described in Table 1.*

In the configurations we use the delimiter ";" instead of "," to avoid confusion with the "," separating the atoms in a goal. We will use the notations $\longrightarrow^*$ to denote the transitive reflexive closure of the transition relation $\longrightarrow$, and $\not\longrightarrow$ to indicate the absence of any further transition.

The transition system models the inherent parallelism of logic programs in terms of processes interacting via a common datastructure, i.e. the accumulated set of bindings. A configuration $\langle G; \vartheta \rangle$ specifies the goal (or network) $G\vartheta$, where G represents the *generic structure* of the process part and ϑ the accumulated set of bindings. This distinction in the configuration is introduced in order to model the communication in the parallel rule **C2**. We call the components of $G\vartheta$ *instances* (of the corresponding components of G), or, more simply, *processes*.

Rule **C1** describes the resolution step of a single process; P_{ren} is the closure of the program with respect to variable renaming. Rule **C2** models the parallel execution of the processes in an *interleaved* manner; additionally it allows for the transmission of the bindings produced by a single process to the common datastructure, hence to the parallel processes. In order to cope properly with the renaming mechanism of logic programming in this parallel setting, we assume that the variables of the renamed clause used for the resolution step do not occur either in $\bar{A}$, $\bar{B}$ or ϑ. The interested reader can find in (Saraswat and Rinard 1990) and in (de Boer and Palamidessi 1991a) formal solutions to this problem. Furthermore we assume that only one possible renaming is considered in **C1**; as a consequence we have finite-branching nondeterminism.

From the above described operational model we can abstract various other semantics, depending on what we want to *observe* of a process. At first we consider a very simple notion of observables, corresponding to the standard operational semantics of logic programming. We associate with each goal the set of substitutions resulting from successfully terminating computations, and *ff* if the process fails. A computation is *fair* if it is either finite or each of the atoms in the configurations is eventually involved in the application of Rule **C1**.

Definition 3.2 (Observables) *Given a program P, for every goal G we define*[11]

$$\mathcal{O}_{ss}(G) = \{\vartheta_{\restriction G} \mid \langle G; \epsilon \rangle \longrightarrow^* \langle \square; \vartheta \rangle \},$$

$$\mathcal{O}_{\mathit{ff}}(G) = \{\mathit{ff} \mid \text{ for all fair computations starting from } \langle G; \epsilon \rangle \\ \text{there exists } G' \text{ s.t. } \langle G; \epsilon \rangle \longrightarrow^* \langle G'; \vartheta \rangle \not\longrightarrow, \\ \text{and } G' \neq \square \text{ holds} \}$$

where $\vartheta_{\restriction G}$ *indicates the restriction of the domain of ϑ to the variables of G.*

Note that $\mathcal{O}_{ss}(G) = \emptyset$ does not imply $\mathit{ff} \in \mathcal{O}_{\mathit{ff}}(G)$; this is due to the possible presence of infinite fair computations. The standard notions of success set and failure set can be retrieved from the above defined observables as follows:

---

[11] Equivalently, $\mathcal{O}_{\mathit{ff}}(G)$ could be defined as a Boolean. We have chosen the set notation for the sake of uniformity with the success case and with more complicated notions of failure which will be considered later.

$$SS = \{A \mid A \text{ is ground and } \epsilon \in \mathcal{O}_{ss}(\leftarrow A)\},$$
$$FF = \{A \mid A \text{ is ground and } f\!f \in \mathcal{O}_{f\!f}(\leftarrow A)\}.$$

Next we show that the success component of the observables is compositional with respect to the parallel operator (i.e. the logical conjunction). In other words, the substitutions resulting from the execution of a network of processes can be obtained by executing the single processes independently, and then combining their results. This should be contrasted with the case of imperative parallel languages, in which compositionality requires information about the intermediate states of the computations, and can be interpreted as a formal justification of the claim that logic programming is inherently parallel. The main reason of this difference is that in logic programming the state (i.e. the accumulated set of bindings) evolves monotonically. In the presence of non-monotonic predicates, like var, this result does not hold for logic programming either, see (Apt et al. 1992).

Compositionality is the basis of a denotational approach to the semantics, which will be treated in Section 5.4. In order to show the compositionality of $\mathcal{O}_{ss}$ it is convenient to consider a *parallel composition* operator on substitutions (Palamidessi 1990; Jacquet 1991), which combines the bindings on shared variables produced independently by parallel processes.

A substitution ϑ is naturally associated with a set of equations

$$\mathcal{E}(\vartheta) = \{x = t \mid x \in dom(\vartheta) \text{ and } \vartheta(x) = t\}.$$

Given a set of equations $E = \{t_1 = u_1, \ldots, t_n = u_n\}$, we extend the notion of most general unifier by defining $mgu(E)$ as the least (up to $\simeq$) substitution ϑ such that for each $i \in [1, n]$ we have $t_i \vartheta = u_i \vartheta$. If such a substitution does not exist then we define $mgu(E) = false$.

Definition 3.3 *The parallel composition of two substitutions ϑ and σ is defined as*

$$\vartheta \parallel \sigma = mgu(\mathcal{E}(\vartheta) \cup \mathcal{E}(\sigma)).$$

Proposition 3.4 (Compositionality of $\mathcal{O}_{ss}$, *(de Boer et al. 1992))*
For any goal $\leftarrow \bar{A}, \bar{B}$

$$\mathcal{O}_{ss}(\leftarrow \bar{A}, \bar{B}) = \mathcal{O}_{ss}(\leftarrow \bar{A}) \;\tilde{\parallel}_{ss}\; \mathcal{O}_{ss}(\leftarrow \bar{B})$$

where for sets of substitutions X and Y

$$X \;\tilde{\parallel}_{ss}\; Y = \{\vartheta \parallel \sigma \mid \vartheta \in X,\ \sigma \in Y \text{ and } \vartheta \parallel \sigma \neq false\}.$$

The simplicity of this result (compositionality of $\mathcal{O}_{ss}$) is due essentially to (a) the already mentioned monotonic evolution of the state, and (b) the absence of any suspension mechanism for process synchronization. In the following sections we will show that such a synchronization mechanism requires more elaborate semantic structures.

Concerning the failure component of the semantics, the following example shows that it is not compositional.

Example 3.5 Consider the program

$$p(a) \leftarrow p(a),$$
$$q(b) \leftarrow q(b),$$
$$r(a) \leftarrow .$$

We have that $\leftarrow p(x)$ and $\leftarrow q(x)$ are observationally equivalent, since $\mathcal{O}_{ss}(\leftarrow p(x)) = \mathcal{O}_{ss}(\leftarrow q(x)) = \emptyset$ and $\mathcal{O}_{ff}(\leftarrow p(x)) = \mathcal{O}_{ff}(\leftarrow q(x)) = \emptyset$. On the other hand, when $p(x)$ and $q(x)$ are inserted in a parallel context containing $r(x)$ the equivalence is lost: $\mathcal{O}_{ff}(\leftarrow p(x), r(x)) = \emptyset$ while $\mathcal{O}_{ff}(\leftarrow q(x), r(x)) = \{ff\}$.

Full compositionality can be achieved by observing also the "limit results" computed by the infinite derivations. In short, the idea is the following: first turn the domain of substitutions into a CPO (by using one of the standard completion techniques), then extend the notions of restriction and *mgu* to the infinite elements (Levi et al. 1990), finally define the infinite component of the observables $\mathcal{O}_{ii}$ as follows:

$$\mathcal{O}_{ii}(G) = \{(lub_n \vartheta_n)_{\restriction G} \mid \text{ there exists an infinite fair computation} \\ \langle G; \epsilon \rangle \longrightarrow \langle G_1; \vartheta_1 \rangle \longrightarrow \ldots \langle G_n; \vartheta_n \rangle \longrightarrow \ldots \}.$$

In the above definition the infinite computations are required to be fair essentially because also the failures are defined in terms of fair computations. Arbitrary infinite computations would not provide the necessary information to model failure compositionally. The reader is invited to find an example of two processes having the same results when arbitrary infinite computations are also considered, and which behave differently with respect to failure when put in parallel with another process.

With this new definition, the observables are fully compositional. For the failure component we have

$$\mathcal{O}_{ff}(\leftarrow \bar{A}, \bar{B}) = \mathcal{O}(\leftarrow \bar{A}) \, \tilde{\|}_{ff} \, \mathcal{O}(\leftarrow \bar{B})$$

where $\mathcal{O}(G) = \mathcal{O}_{ss}(G) \cup \mathcal{O}_{ff}(G) \cup \mathcal{O}_{ii}(G)$ and

$$X \, \tilde{\|}_{ff} \, Y \;=\; \{ff \mid \; ff \in X \cup Y \text{ or} \\ \forall \vartheta \in X \; \forall \sigma \in Y \; \vartheta \parallel \sigma = false\}.$$

For the infinite component:

$$\mathcal{O}_{ii}(\leftarrow \bar{A}, \bar{B}) = \langle \mathcal{O}_{ss}(\leftarrow \bar{A}), \mathcal{O}_{ii}(\leftarrow \bar{A}) \rangle \, \tilde{\|}_{ii} \, \langle \mathcal{O}_{ss}(\leftarrow \bar{B}), \mathcal{O}_{ii}(\leftarrow \bar{B}) \rangle,$$

where

$$\langle X_1, X_2 \rangle \, \tilde{\|}_{ii} \, \langle Y_1, Y_2 \rangle \;=\; \{\vartheta \,\|\, \sigma \mid \vartheta \in X_2, \sigma \in Y_1 \cup Y_2$$
$$\text{and } \vartheta \,\|\, \sigma \neq \textit{false} \,\}$$
$$\cup \;\; \{\vartheta \,\|\, \sigma \mid \vartheta \in X_1 \cup X_2, \sigma \in Y_2$$
$$\text{and } \vartheta \,\|\, \sigma \neq \textit{false} \,\}.$$

Example 3.6 For the program in Example 3.5 we have $\mathcal{O}_{ii}(\leftarrow p(x)) = \{\{x/a\}\}$ whereas $\mathcal{O}_{ii}(\leftarrow q(x)) = \{\{x/b\}\}$.

3.2 Explicit mechanisms for concurrency

We describe now the main characteristics of concurrent logic languages.

3.2.1 Don't care nondeterminism

The kind of nondeterminism in logic programming is *don't know*, as in (nondeterministic) Turing machines and finite automata, i.e. it is meant for exhaustive search algorithms. Failure is *universal*, namely failing computations are discarded if there are successful ones. Therefore, the implementation of don't know nondeterminism requires some backtracking mechanism.

Concurrent systems imply a different notion: nondeterminism is needed for coping with unpredictable circumstances, like the relative speed of the various processes, and the interaction with the external world. A concurrent process is required to select one of the alternatives which are enabled at a certain moment, and to *commit* to that choice. Note that this induces an *existential* notion of failure. In any case, it seems not to make much sense for a process to backtrack, because it would involve undoing actions that have already influenced the evolution of the environment. This kind of nondeterminism is called *don't care*.

From a formal point of view, the computational model of logic programming with don't care nondeterminism is still expressed by the transition system described in Table 1. The notion of successful computation remains the same. What changes is the notion of failure, which becomes existential:

$$\mathcal{O}_{f\!f}(G) \;=\; \{f\!f \mid \text{ there exist } G', \vartheta \text{ such that }$$
$$\langle G; \epsilon \rangle \longrightarrow^* \langle G'; \vartheta \rangle \not\longrightarrow \text{ and } G' \neq \square \}.$$

Note that with don't care nondeterminism the presence of $f\!f$ in the semantics of a process does not exclude successful or infinite computations.

This notion of failure is not compositional. The reason is that existential failure depends on the *branching structure* of a process, namely the tree which represents the alternative computations of a process, where a choice point indicates the selection of an atom and of a clause for the resolution step.

Example 3.7 Consider the program

$$p(a) \leftarrow,$$
$$p(b) \leftarrow,$$
$$q(x) \leftarrow r_1(x),$$
$$q(x) \leftarrow r_2(x),$$
$$r_1(a) \leftarrow,$$
$$r_2(b) \leftarrow.$$

Consider the network $\leftarrow p(x), p(x)$. Whatever binding ($\{x/a\}$ or $\{x/b\}$) is generated by one of the two p's processes, the other one will select the right clause accordingly. Therefore failure is not possible: $\mathcal{O}_{\!f\!f}(\leftarrow p(x), p(x)) = \emptyset$. Consider now the network $\leftarrow p(x), q(x)$. The choice of the clause for q is *blind*, hence after p has produced a binding, say, $\{x/a\}$, q can still make the wrong choice by selecting the clause $q(x) \leftarrow r_2(x)$, which leads to failure. Therefore $\mathcal{O}_{\!f\!f}(\leftarrow p(x), q(x)) = \{f\!f\}$.

The difference in the choice structure between $\leftarrow p(x)$ and $\leftarrow q(x)$ is not reflected by the observables. In fact $\leftarrow p(x)$ and $\leftarrow q(x)$ are observationally equivalent, since $\mathcal{O}_{ss}(\leftarrow p(x)) = \mathcal{O}_{ss}(\leftarrow q(x)) = \{\{x/a\}, \{x/b\}\}$, $\mathcal{O}_{\!f\!f}(\leftarrow p(x)) = \mathcal{O}_{\!f\!f}(\leftarrow q(x)) = \emptyset$ and $\mathcal{O}_{ii}(\leftarrow p(x)) = \mathcal{O}_{ii}(\leftarrow q(x)) = \emptyset$.

In Section 5.4 we will see how to encode the relevant information of the branching structure so as to achieve compositionality.

Since the main feature of reactive systems is the possibly ongoing interaction with the environment, the infinite behavior is of interest in its own right.

$$\mathcal{O}_{ii}(G) = \{(lub_n \vartheta_n)_{\restriction G} \mid \text{ there exists an infinite computation}$$
$$\langle G; \epsilon \rangle \longrightarrow \langle G_1; \vartheta_1 \rangle \longrightarrow \ldots \langle G_n; \vartheta_n \rangle \longrightarrow \ldots \}.$$

3.2.2 *Control of nondeterminism*

Don't care nondeterminism is usually combined with the notion of *guard* (Dijkstra 1986), which allows control of the choices. In general, guards are conditions associated with the various alternatives in a choice construct. An alternative can be selected (is enabled) only if the associated guard is satisfied. In other words, it is a sort of conditional statement. In a concurrent setting, the guarded choice is often combined with a synchronization mechanism, i.e. the conditions are influenced by the other processes.

The introduction of guards in logic programming leads to the concept of *guarded clause*, which is an expression of the form

$$A \leftarrow \bar{g} \mid \bar{B},$$

Concurrent logic and concurrent constraint programming

$$\mathbf{C1} \quad \langle \leftarrow A; \vartheta \rangle \longrightarrow \langle \leftarrow \bar{B}; \vartheta\sigma\eta \rangle \qquad \begin{array}{l} A' \leftarrow \bar{g} \mid \bar{B} \in P_{ren}, \\ \sigma = mgu(A\vartheta, A') \text{ and} \\ eval(\bar{g}\sigma) = true \\ \text{with substitution } \eta \end{array}$$

Table 2 The resolution rule for flat guards

where $\bar{g}$ is the guard, consisting of a sequence of atoms; "|" is the *commit* operator, and $\bar{B}$ is the body.

The guard $\bar{g}$ can contain built-in and user-defined predicates. In case there are only built-in's, $\bar{g}$ is *flat*, otherwise it is *deep*.

Built-in predicates are for instance the PROLOG built-in's. However, we abstract here from their specific nature, and we only assume the presence of some mechanism *eval* embedded in the language which evaluates each sequence of built-in atoms $\bar{C}$ either to *true* or to *false* and in the first case it delivers a (possibly empty) substitution.

In the flat case, the operational model of guarded clauses can be obtained by a simple modification of the system described in Table 1: we only need to add in Rule **C1** the condition on the guard. The new rule is described in Table 2. The semantics $\mathcal{O}_{ss}$, $\mathcal{O}_{ff}$, and $\mathcal{O}_{ii}$ are defined as before ($\mathcal{O}_{ff}$ being of course the one specified for don't care nondeterminism).

Let's now consider deep guards. A first idea would be to replace the condition $eval(\bar{g}\sigma) = true$ *with substitution* η in **C1** by the premise $\langle \leftarrow \bar{g}\sigma; \epsilon \rangle \longrightarrow^* \langle \Box; \eta \rangle$. This corresponds to the intended meaning of deep guards, and it is sufficient to describe successful computations. On the other hand, when we consider failure, things become more complicated. Concurrent logic languages with deep guards adopt a "deep" notion of failure, i.e. failure can be caused also by a failure in the guard evaluation.

Example 3.8 Consider the program

$$p \leftarrow q \mid,$$
$$q \leftarrow \mid,$$
$$q \leftarrow \mid r.$$

In the intended model $\leftarrow p$ has the possibility to fail, which happens when the resolution of the guard $\leftarrow q$ commits to the third clause. This failure is however not captured by the semantics described above: the configuration $\langle \leftarrow p; \epsilon \rangle$ is not stuck since the guard has also the possibility to succeed.

An analogous problem arises for infinite computations.

C1	$\langle \leftarrow A; \vartheta \rangle \longrightarrow \langle \Box; \vartheta\eta \rangle$	A is a built-in and $eval(A\vartheta) = true$ with substitution η
C2	$\dfrac{\langle \leftarrow \bar{g}\sigma; \epsilon \rangle \longrightarrow^* \langle \Box; \eta \rangle}{\langle \leftarrow A; \vartheta \rangle \longrightarrow \langle \leftarrow \bar{B}; \vartheta\sigma\eta \rangle}$	there exists A' s.t. $A' \leftarrow \bar{g} \mid \bar{B} \in P_{ren}$ and $\sigma = mgu(A\vartheta, A')$
C3	$\dfrac{\langle \leftarrow \bar{A}; \vartheta \rangle \longrightarrow \langle \leftarrow \bar{A}'; \vartheta' \rangle}{\langle \leftarrow \bar{A}, \bar{B}; \vartheta \rangle \longrightarrow \langle \leftarrow \bar{A}', \bar{B}; \vartheta' \rangle}$ $\dfrac{}{\langle \leftarrow \bar{B}, \bar{A}; \vartheta \rangle \longrightarrow \langle \leftarrow \bar{B}, \bar{A}'; \vartheta' \rangle}$	

Table 3 Deep guards: computation rules

Example 3.9 Consider the programs

$$P_1 \quad p \leftarrow q \mid,$$
$$q \leftarrow q \mid,$$
$$P_2 \quad p \leftarrow q \mid,$$
$$q \leftarrow \mid,$$
$$q \leftarrow q \mid.$$

In P_1 the evaluation of the guard $\leftarrow q$ diverges; in the semantics described above $\langle \leftarrow p; \epsilon \rangle$ has no outgoing transitions. In P_2 the guard $\leftarrow q$ can still diverge, but this possibility is not reflected in the above semantics: $\langle \leftarrow p; \epsilon \rangle$ has an outgoing transitions corresponding to the successful derivation of $\langle \leftarrow q; \epsilon \rangle$ by using the second clause.

The above examples show that the definition of $\mathcal{O}_{ff}$ and $\mathcal{O}_{ii}$ require a much more elaborate formulation. However for the case of failure we can obtain a concise description by adding some additional transition rules. In Tables 3 and 4 we give the rules for success and failure; for a treatment of divergence we refer to (Knijnenburg and Kok 1989). For further discussion on the operational semantics of deep guards see (Saraswat 1987a) and (Levy 1988).

F1	$\langle \leftarrow A; \vartheta \rangle \longrightarrow \langle f\!f; \vartheta \rangle$	A is a built-in and $eval(A\vartheta) = false$
F2	$\langle \leftarrow A; \vartheta \rangle \longrightarrow \langle f\!f; \vartheta \rangle$	for all $A' \leftarrow \bar{g} \mid \bar{B} \in P_{ren}$, $mgu(A\vartheta, A') = false$ or $\langle \leftarrow \bar{g}\ mgu(A\vartheta, A'); \epsilon \rangle \longrightarrow^* \langle f\!f; \eta \rangle$ for some η
F3	$\dfrac{\langle \leftarrow \bar{A}; \vartheta \rangle \longrightarrow \langle f\!f; \eta \rangle}{\langle \leftarrow \bar{A}, \bar{B}; \vartheta \rangle \longrightarrow \langle f\!f; \eta \rangle}$ $\langle \leftarrow \bar{B}, \bar{A}; \vartheta \rangle \longrightarrow \langle f\!f; \eta \rangle$	

Table 4 Deep guards: failure rules

The observables $\mathcal{O}_{ss}$ are defined as usual; for $\mathcal{O}_{f\!f}$ we have

$$\mathcal{O}_{f\!f}(G) = \{f\!f \mid \text{there exists } \vartheta \text{ such that } \langle G; \epsilon \rangle \longrightarrow^* \langle f\!f; \vartheta \rangle\}.$$

Most of the deep concurrent logic languages adopt the principle that guards are evaluated in a *fair manner*. Roughly, this means that in the derivation step all guards are tried in parallel; if one guard is successful and the others are not then the former is eventually selected. The following is one of the typical examples used by Shapiro (1983) to illustrate the usefulness of fair evaluation of guards.

Example 3.10 (**Interrupt-handler in a Unix-like shell**) Consider an interactive system consisting of a process *user* and a process *shell* (see Figure 1). The user sends a stream (i.e. a possibly infinite list) l of commands to the shell, and the shell executes them. If the stream contains an *interrupt*, namely a command of the form $"\!\uparrow c"$ (control_c), then the shell must abort the current execution, flush the portion of stream up to the element $"\!\uparrow c"$, and resume the execution from there.

Using the notation $[x \mid l]$ to represent the list whose head is x and whose tail is l, the program for shell can be written as follows:

$$shell([x \mid l]) \leftarrow command(x) \mid shell(l),$$

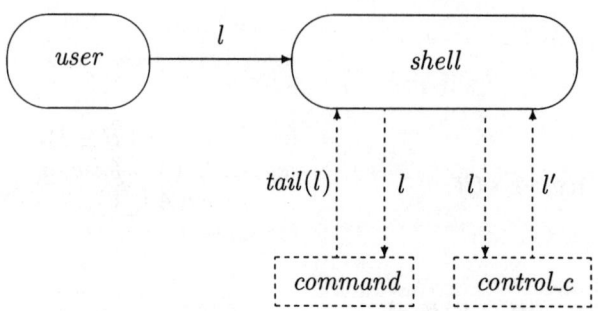

FIG. 1. Interrupt-handler in a Unix-like shell.

$shell(l) \leftarrow control\_c(l, l') \mid shell(l'),$

$control\_c([x \mid l], l') \leftarrow x \neq ''\uparrow c'' \mid control\_c(l, l')\},$

$control\_c([''\uparrow c'' \mid l], l') \leftarrow \mid l = l',$

...

where the process $command(x)$ is supposed to execute the command x. The goal $\leftarrow control\_c(l, l')$ searches for an occurrence of $''\uparrow c''$ in l; in case of success, it produces the stream l' of commands following $''\uparrow c''$ in l. A call to $command(x)$ may cause an infinite loop; in this case the fairness of the guard evaluation ensures that a possible interrupt sent by the user will eventually be effective.

For further discussion about fairness in concurrent logic programming, and for an interesting idea about the possibility of replacing parallel guard evaluation by AND parallelism, see (Saraswat 1986).

3.2.3 Deep guards versus flat guards

Deep guards are clearly more expressive than flat guards. In particular, they permit the implementation of *unbounded nondeterminism*, i.e. choice constructs with infinitely many alternatives. In other words, deep guards allow infinite-branching nondeterminism, in contrast with the finite-branching nondeterminism of flat guards.

Example 3.11 Consider the program

$$p(s(x)) \leftarrow p(x) \mid,$$
$$p(0) \leftarrow \mid.$$

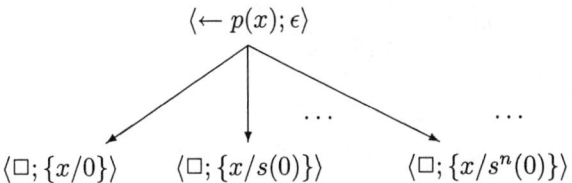

FIG. 2. Transitions in the program with deep guards.

From the configuration $\langle \leftarrow p(x); \epsilon \rangle$ infinitely many transitions are possible, see Figure 2.

Note that this process cannot be specified in a language with finite-branching nondeterminism. In fact the transition graph would be a finite-branching tree with infinitely many leaves, hence by König's lemma there would be an infinite derivation. For instance, the following flat version of the program

$$p(s(x)) \leftarrow\mid p(x),$$
$$p(0) \leftarrow\mid,$$

gives rise to the transitions shown in Figure 3.

However, apart from infinite-branching nondeterminism, in most cases deep guards are not necessary. Furthermore, deep guards are quite difficult to implement, especially in the presence of synchronization, and particularly when atomic unification (see Section 3.3.4) is adopted; see (Takeuchi and Furukawa 1986) for a discussion about this topic. For these reasons, nowadays deep guards are not much used anymore. Perhaps the only example of a language in which they are still used is Andorra PROLOG (Haridi and Brand 1988; Haridi 1990; Haridi and Janson 1990, 1991; Haridi et al. 1992), where they are used in combination with don't know nondeterminism.

3.2.4 Synchronization

Most of the concurrent logic languages realize synchronization by means of restrictions on the resolution step (head unification, guard evaluation). Usually these restrictions depend on the state of the variables, which is influenced by the actions done by the other processes in the network.

In the presence of this synchronization mechanism, a process can get stuck not only because of the absence of compatible clauses (failure), but also because the conditions which control its next steps are not satisfied. In this case the process *is suspended*. Later on, the bindings on common

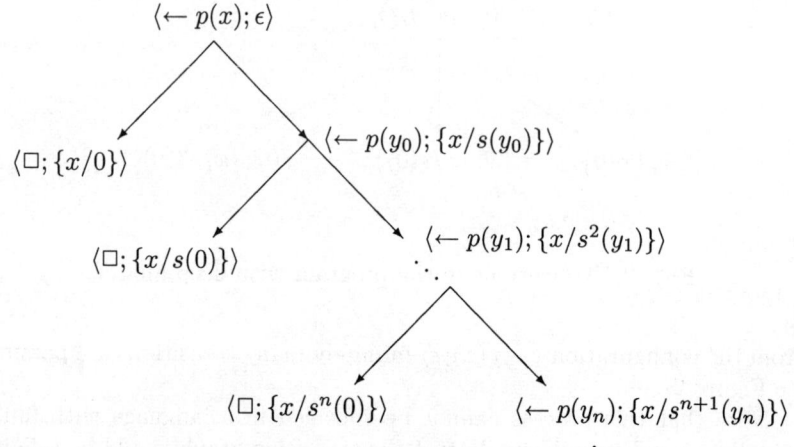

FIG. 3. Transitions in the program with flat guards. For simplicity, only the bindings on x are represented.

variables established by the other processes might cause the conditions to get satisfied, and in this case the process is *resumed*. Thus suspension is not definitive, while failure is. Another difference is that failure is *catastrophic*, namely when a process fails the whole network fails, whereas the network suspends (deadlocks) only if all the processes suspend. For these reasons usually one distinguishes in the observables these two kinds of termination mode. We use an additional symbol, dd, to represent suspension.

In order to describe this synchronization mechanism in a general way we assume two conditions, $cond_1$ and $cond_2$, associated with the head unification and with the guard evaluation respectively, which have to be tested before committing to a certain clause. Furthermore we assume a third condition, $cond_3$, specific for built-in's. These conditions depend on the binding which would be generated and on the selected atom. Tables 5-7 specify the transition system for a language with flat guards. Deep guards are more complicated; the reader interested in a formal treatment of them is referred to (Saraswat 1987a) and to (Levy 1988).

The observables are defined as follows:

Definition 3.12 (Observables) *Given a program P, for every goal G*

C1	$\langle \leftarrow A; \vartheta \rangle \longrightarrow \langle \Box; \vartheta\eta \rangle$	A is a built-in and $eval(A\vartheta) = true$ with substitution η and $cond_3(\eta, A\vartheta)$ holds
C2	$\langle \leftarrow A; \vartheta \rangle \longrightarrow \langle \bar{B}; \vartheta\sigma\eta \rangle$	there exists A' s.t. $A' \leftarrow \bar{g} \mid \bar{B} \in P_{ren}$, $\sigma = mgu(A\vartheta, A')$, $eval(\bar{g}\sigma) = true$ with substitution η and $cond_1(\sigma, A\vartheta)$, $cond_2(\sigma\eta, A\vartheta)$ hold
C3	$\dfrac{\langle \leftarrow \bar{A}; \vartheta \rangle \longrightarrow \langle \leftarrow \bar{A}'; \vartheta' \rangle}{\langle \leftarrow \bar{A}, \bar{B}; \vartheta \rangle \longrightarrow \langle \leftarrow \bar{A}', \bar{B}; \vartheta' \rangle}$ $\langle \leftarrow \bar{B}, \bar{A}; \vartheta \rangle \longrightarrow \langle \leftarrow \bar{B}, \bar{A}'; \vartheta' \rangle$	

Table 5 Synchronization and flat guards: computation rules

F1	$\langle \leftarrow A; \vartheta \rangle \longrightarrow \langle f\!\!f; \vartheta \rangle$	A is a built-in and $eval(A\vartheta\sigma) = false$
F2	$\langle \leftarrow A; \vartheta \rangle \longrightarrow \langle f\!\!f; \vartheta \rangle$	for all $A' \leftarrow \bar{g} \mid \bar{B} \in P_{ren}$, $mgu(A\vartheta, A') = false$ or $eval(\bar{g}\, mgu(A\vartheta, A')) = false$
F3	$\dfrac{\langle \leftarrow \bar{A}; \vartheta \rangle \longrightarrow \langle f\!\!f; \eta \rangle}{\langle \leftarrow \bar{A}, \bar{B}; \vartheta \rangle \longrightarrow \langle f\!\!f; \eta \rangle}$ $\langle \leftarrow \bar{B}, \bar{A}; \vartheta \rangle \longrightarrow \langle f\!\!f; \eta \rangle$	

Table 6 Synchronization and flat guards: failure rules

S1	$\langle \leftarrow A; \vartheta \rangle \longrightarrow \langle dd; \vartheta \rangle$	A is a built-in and $eval(A\vartheta) = true$ with substitution η and $cond_3(\eta, A\vartheta)$ does not hold
S2	$\langle \leftarrow A; \vartheta \rangle \longrightarrow \langle dd; \vartheta \rangle$	Rules **C2** and **F2** do not apply
S3	$\dfrac{\langle \leftarrow \bar{A}; \vartheta \rangle \longrightarrow \langle dd; \vartheta \rangle \quad \langle \leftarrow \bar{B}; \vartheta \rangle \longrightarrow \langle dd; \vartheta \rangle}{\langle \leftarrow \bar{A}, \bar{B}; \vartheta \rangle \longrightarrow \langle dd; \vartheta \rangle}$	

Table 7 Synchronization and flat guards: suspension rules

we define

$$\mathcal{O}_{ss}(G) = \{\vartheta_{\restriction G} \mid \langle G; \epsilon \rangle \longrightarrow^* \langle \square; \vartheta \rangle\},$$

$$\mathcal{O}_{\mathit{ff}}(G) = \{\mathit{ff} \mid \text{there exists } \vartheta \text{ such that } \langle G; \epsilon \rangle \longrightarrow^* \langle \mathit{ff}; \vartheta \rangle\},$$

$$\mathcal{O}_{dd}(G) = \{\vartheta_{\restriction G} \mid \langle G; \epsilon \rangle \longrightarrow^* \langle dd; \vartheta \rangle\},$$

$$\mathcal{O}_{ii}(G) = \{(lub_n \vartheta_n)_{\restriction G} \mid \text{there exists an infinite computation } \langle G; \epsilon \rangle \longrightarrow \langle G_1; \vartheta_1 \rangle \longrightarrow \ldots \langle G_n; \vartheta_n \rangle \longrightarrow \ldots\}.$$

Note that the failure rules (**F1-F3**) and the suspension rules (**S1-S3**) are complementary and that in order to describe the observables it is not necessary to have both sets of rules. For instance, we could remove Rules **S1-S3** and the configurations of the form $\langle dd, \vartheta \rangle$, and describe $\mathcal{O}_{dd}$ in terms of the absence of outgoing transitions.

3.3 Variables as input output channels

In most languages the conditions for synchronization specify that certain variables may not get instantiated during the resolution step. These specific conditions give rise to a classification of the variables into *input* and *output* variables, the former being the ones to which the above restriction applies. This leads to the concept of a system of *producers* and *consumers*; the *flow of data* from producers to consumers requires the consumer to wait until some specified data has been produced.

In this section we give an overview of the main languages which are

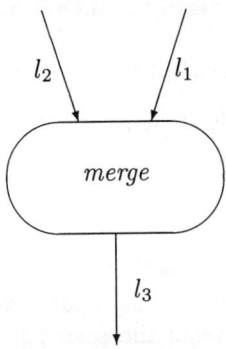

FIG. 4. The merge process.

based on this kind of synchronization mechanism.

3.3.1 The Relational Language and PARLOG

In PARLOG (Clark and Gregory 1986; Gregory 1987) input and output variables are specified via a *mode declaration*. For every predicate p there exists a declaration of the form

$$p(\ldots,?,\ldots,\hat{}\,,\ldots),$$

where "?" and "^" indicate the input and output arguments of p, respectively. Thus the condition $cond_1(\vartheta, p(\bar{t}))$ holds iff ϑ does not bind the input arguments of p. A guard is not allowed to bind the input variables; such a situation generates an error.

The Relational Language (Clark and Gregory 1981), which is the predecessor of PARLOG and corresponds to the subset called Directional PARLOG (Gregory 1987), has more restrictions. In particular, the substitution generated by the head unification cannot bind the variables occurring in an output argument of the head of the clause. This means that the output variables of a process can be bound only by the process itself. Channels are strictly unidirectional; multiproducers are not allowed and unification is in practice replaced by (one-way) pattern matching.

As an example we give the specification of the *angelic merge* process in PARLOG. A merge process takes two input streams l_1, l_2 of data and merges them into an output stream l_3, see Figure 4.

Such a process is called angelic if it is guaranteed that when no data is available on one input then all data from the other input will be processed. In the following, [ ] represents the empty list, and signals the *end of the stream*. This must not be confused with the *empty stream*, represented by an unbound list variable.

The angelic merge process in PARLOG can be defined as follows

$$merge([x|l_1], l_2, [x|l_3]) \leftarrow \ |\ merge(l_1, l_2, l_3),$$
$$merge(l_1, [x|l_2], [x|l_3]) \leftarrow \ |\ merge(l_1, l_2, l_3),$$
$$merge([\,], l, l) \leftarrow \ |,$$
$$merge(l, [\,], l) \leftarrow \ |,$$

with the mode declaration $merge(?, ?, \hat{\ })$. The merge specified by the program above is angelic because if one input stream is empty, then the synchronization condition prevents the selection of the clause which would process that input stream. For instance, if the first argument is not bound then only the second or the fourth clause can be selected.

3.3.2 Concurrent PROLOG

In Concurrent PROLOG (Shapiro 1983, 1986, 1988), a variable occurrence in the goal, in the guard or in the body may be decorated with the annotation "?", which marks it as read-only. The conditions $cond_1(\vartheta, A)$ and $cond_2(\vartheta, A)$ are verified iff ϑ does not bind the read-only variables of A to non-variable terms. Variables which are bound to read-only variables inherit the read-only annotation. For example, given a clause $p(x) \leftarrow q(x)$, the goal $\leftarrow p(y?), r(y)$ can evolve in $\leftarrow q(y?), r(y)$.

The *angelic merge* process is specified in Concurrent PROLOG as follows:

$$merge([x|l_1], l_2, [x|l_3]) \leftarrow \ |\ merge(l_1?, l_2, l_3),$$
$$merge(l_1, [x|l_2], [x|l_3]) \leftarrow \ |\ merge(l_1, l_2?, l_3),$$
$$merge([\,], l, l) \leftarrow \ |,$$
$$merge(l, [\,], l) \leftarrow \ |.$$

A call to the merge process should be of the form $\leftarrow merge(l_1?, l_2?, l_3)$.

3.3.3 Guarded Horn Clauses

In the Guarded Horn Clauses language (Ueda 1987, 1988), GHC for short, there is a predefined binary predicate $=$ which performs the unification of the two arguments. The atoms of the form $t = u$ are the only ones which can produce bindings; for all the other atoms A the conditions $cond_1(\vartheta, A)$ and $cond_2(\vartheta, A)$ are validated iff ϑ does not bind any variable of A. In other words, all the variables of A are treated as input variables. Note that therefore the predicate $=$ cannot be defined in GHC.

The angelic merge process can be specified in GHC as follows:

$$merge([x|l_1], l_2, l_3) \leftarrow | l_3 = [x|l'_3], merge(l_1, l_2, l'_3),$$
$$merge(l_1, [x|l_2], l_3) \leftarrow | l_3 = [x|l'_3], merge(l_1, l_2, l'_3),$$
$$merge([\,], l_2, l_3) \leftarrow | l_2 = l_3,$$
$$merge(l_1, [\,], l_3) \leftarrow | l_1 = l_3.$$

3.3.4 *Atomic versus eventual unification*

The main distinguishing feature of Concurrent PROLOG with respect to PARLOG and GHC is the presence of the so-called *atomic unification*. Roughly, this means that the head unification and the guard evaluation can produce bindings (on the variables which are not read-only) and that this is done atomically. In GHC and PARLOG bindings can be produced only after the commitment to a certain clause (*eventual unification*). In GHC the conditions $cond_1$ and $cond_2$ are such that no bindings are produced during the selection of the clause in Rule **C2** of Table 5. In PARLOG, at least in the latter versions, there are some additional restrictions that prevent the production of bindings during the head unification and the guard evaluation.

Our transition system models a *global memory*, represented by the accumulated bindings in the configuration. Furthermore, the head unification and the guard evaluation are seen as one atomic action, i.e. they are performed in one single step. In a different model, either based on *distributed memory*, or where the head unification and the guard evaluation might require more than one step, additional machinery would be necessary to realize the synchronization mechanism of Concurrent PROLOG.

In a system where the head unification and the guard evaluation are not atomic actions, Concurrent PROLOG can be modeled by encapsulating them into a critical section (Knijnenburg and Kok 1989).

On the other hand, a distributed model can be obtained by associating to each process its local memory, where the bindings established by a process are stored, instead of being immediately communicated to the other processes. Of course, the "atomic unification" principle requires a *consistency check* before the commitment, to verify that the bindings produced by the head unification and by the guard evaluation are consistent with the bindings established by the other processes. When a process starts performing the consistency check, the other processes must wait before producing new bindings until it has finished.

Of course, in a distributed model atomic unification is very expensive to implement: the consistency check is a bottleneck which actually reduces the potential parallelism of the system. However, it is very powerful. In fact it can be argued that languages with atomic unification are strictly more expressive than languages without it (de Boer and Palamidessi 1994).

Example 3.13 Suppose we want to define a process $p(x)$ which produces either $\{x/a\}$ or $\{x/b\}$, and that maintains these possibilities (without failing) when put in parallel with itself. In Concurrent PROLOG such $p(x)$ can be specified by the program:

$$p(a) \leftarrow |,$$
$$p(b) \leftarrow | \,.$$

On the other hand this process cannot be specified in GHC or in PARLOG. The "best approximation" in GHC would be the program

$$p(x) \leftarrow | \; x = a,$$
$$p(x) \leftarrow | \; x = b,$$

but such a program does not satisfy all the requirements, since the goal $\leftarrow p(x), p(x)$ might fail.

A more interesting example is shown by Shapiro (1989). Such example, taken from a protocol proposed by Safra (1986), can be illustrated as follows. There are two "boy" processes and two "girl" processes. Boys and girls are connected by one-to-one channels. Each boy wishes to "start a relation" with one girl, and he doesn't care who. Each girl wishes to start a relation with one boy, and she doesn't care who. We want to program the nodes so to establish the couples, without risk of deadlock. The solution proposed in (Safra 1986; Shapiro 1989) is relative to the case in which boys take the initiative. Each boy sends to the girls the incomplete message $hello(x)$ (where x is a local variable). Each girl selects nondeterministically one message and instantiates x to her unique identifier. The program is as follows:

$$boy(x, y_1, y_2) \quad \leftarrow \quad | \; y_1 = hello(x), y_2 = hello(x),$$

$$girl(Id, hello(x_1), z) \quad \leftarrow \quad x_1 = Id \;|,$$
$$girl(Id, y, hello(x_2)) \quad \leftarrow \quad x_2 = Id \;| \,.$$

The initial network, also illustrated in Figure 5, is

$$\leftarrow boy(x_1, y_1, y_2), boy(x_2, z_1, z_2), girl(1, y_1?, z_1?), girl(2, y_2?, z_2?).$$

This example shows that atomic unification allows mimicking the communication protocol of synchronous concurrent languages like CSP, where it is possible to define processes whose choices depend upon the outcomes of simultaneous interactions with the environment. Thus we can say in a sense

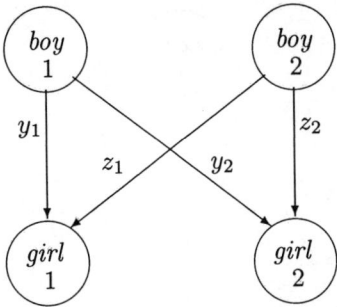

FIG. 5. Partners selection.

that atomic unification makes asynchronous communication as powerful as synchronous communication.

The case in which both boys and girls take the initiative is more complicated, but still solvable with atomic unification. Particularly interesting is the case of a symmetric network, in which all nodes are copies of the same process, i.e. they are defined by the same program. The interested reader is invited to write such program in Concurrent PROLOG. This case seems very difficult to describe in a language with eventual unification. Our conjecture is that it is not possible at all.

Symmetricity seems to be the key point in which atomic unification shows its supremacy concerning expressivity. Let's consider the problem of finding *symmetric algorithms for the election of a leader in a symmetric network*. Such problem was analyzed by Bougé (1988) to show that CSP without output guards is strictly less expressive than full CSP (Hoare 1978). In fact, it is possible to write a program deadlock-free in full CSP which guarantees that a leader is eventually elected, whereas this is not possible in the sublanguage without output guards. The proof of Bougé can be adapted to show that an algorithm satisfying these requirements cannot be written in a concurrent logic language with eventual unification. On the other hand, it can be written by using atomic unification:

$$node(x, y, Id) \leftarrow x = elected(Id), y = elected(Id) \mid ,$$
$$node(x, y, Id) \leftarrow x? = elected(w) \mid ,$$
$$node(x, y, Id) \leftarrow y? = elected(w) \mid .$$

The initial network, which is illustrated in Figure 6, is

$$\leftarrow node(x, y, 1), node(y, z, 2), node(z, x, 3).$$

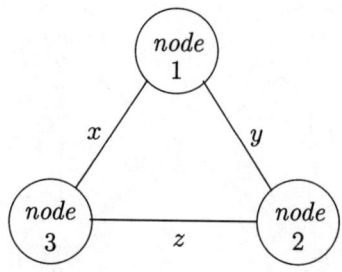

FIG. 6. Election of a leader.

C2	$\langle \leftarrow A; \vartheta \rangle \longrightarrow \langle \bar{B}; \vartheta\sigma \rangle$	there exists $C = A' \leftarrow \bar{g} \mid \bar{B} \in P_{ren}$ s.t. $try(A\vartheta, C) = \sigma$ and for all $C' \in P_{ren}$ if $C' \neq C$ then $try(A\vartheta, C') = \mathit{ff}$

Table 8 The computation rule of P-PROLOG

3.4 An alternative approach to synchronization: P-PROLOG

The synchronization mechanism of P-PROLOG (Yang 1986; Yang and Aiso 1986) is completely different from those of the languages considered so far. The basic idea is that a process can proceed only when there is only one clause which can be applied for the resolution step. Formally this mechanism is described as follows. Given an atom A and a clause $C = A' \leftarrow \bar{g} \mid \bar{B}$, let $try(A, C)$ be defined by

$$try(A, C) = \begin{cases} \sigma\eta & \text{if mgu}(A, A') = \sigma \text{ and } \langle \leftarrow \bar{g}\sigma; \epsilon \rangle \longrightarrow^* \langle \square; \eta \rangle \\ \mathit{ff} & \text{if mgu}(A, A') = \mathit{false} \text{ or } \\ & \text{mgu}(A, A') = \sigma \text{ and } \exists \eta. \langle \leftarrow \bar{g}\sigma; \epsilon \rangle \longrightarrow^* \langle \mathit{ff}; \eta \rangle \\ \mathit{dd} & \text{otherwise} \end{cases}$$

The transition system of P-PROLOG is obtained by modifying the Rule **C2** of Table 5 as described in Table 8.

Note that P-PROLOG is a deterministic language, in the following sense:

Proposition 3.14 *Let P be a P-PROLOG program and $\leftarrow \bar{A}$ be a P-PROLOG goal. If $\langle \leftarrow \bar{A}; \epsilon \rangle$ has a successful computation in P with c.a.s. ϑ, then every computation of $\langle \leftarrow \bar{A}; \epsilon \rangle$ in P terminates with success and a c.a.s. equivalent (up to renaming) to ϑ.*

As a consequence, typical non-deterministic processes, such as the merge process, cannot be expressed in P-PROLOG.

The main use of this synchronization mechanism is to guarantee the correct "flow of data" among the parallel components of a network, thus avoiding redundant or wrong computations.

Example 3.15 Consider the following program, which computes the Fibonacci function

$$fib(0, s(0)) \leftarrow |,$$

$$fib(s(0), s(0)) \leftarrow |,$$

$$fib(s(s(x)), y) \leftarrow | \; fib(s(x), z), fib(s(x), w), plus(z, w, y),$$

$$plus(0, y, y) \leftarrow |,$$

$$plus(s(x), y, s(z)) \leftarrow | \; plus(x, y, z).$$

Consider the initial goal $\leftarrow fib(s(s(0)), y)$. After one resolution step, the goal becomes $\leftarrow fib(s(0), z), fib(0, w), plus(z, w, y)$. Now, the *plus* process cannot proceed until the first process *fib* has computed a binding for z.

The principle of P-PROLOG has been embodied later on in the ALPS languages and in Andorra-PROLOG.

4 The ALPS languages

Recently Jaffar and Lassez (1987) proposed the paradigm of Constraint Logic Programming (CLP), an extension to logic programming which gained immediate popularity in the community. The extension consists of considering an arbitrary domain of data instead of the standard Herbrand universe, and by replacing the concept of unification with the concept of constraint over such domain. The unifiability test becomes a test of consistency of the new constraint with respect to the constraints previously established.

More formally, a CLP clause and goal have the form

$$A \leftarrow \bar{c}, \bar{B},$$

$$\leftarrow \bar{c}, \bar{A},$$

where $\bar{c}$ is a sequence of constraints. The computation step is defined as follows: let A_i be an atom in the goal

$$\leftarrow \bar{c}', \bar{A},$$

and let $H \leftarrow \bar{c}, \bar{B}$ be a renamed clause of the program, such that $\bar{c} \wedge A_i = H \wedge \bar{c}'$ is satisfiable (consistent), where "$\wedge$" is the logical conjunction. Then the new goal becomes

$$\leftarrow \bar{c}, \bar{c}', A_i = H, \bar{A} \setminus A_i, \bar{B}.$$

Maher (1987) has developed a concurrent version of CLP, which he called "The ALPS family" (each member being associated to a constraint system). Among the other features, his proposal contains an innovative principle for synchronization: the *validation rule*. From a technical point of view this rule is the natural extension of the GHC synchronization mechanism to the constraint framework, but the very elegant, logical formulation has induced the logic programming community to regard this idea as the "right" approach to synchronization. Also the *ask rule* of ccp is essentially based on the same concept.

In order to describe formally the ALPS family we briefly survey the notion of constraint system. We follow here the approach of Saraswat and Rinard (1990), which is based on Scott's theory of *information systems* (Scott 1982).

4.1 Constraint systems

Saraswat and Rinard (1990) define a constraint system starting from a primitive notion of simple constraints (tokens) on which a compact entailment relation $\vdash$ is defined. Then a constraint system is constructed by considering sets of tokens and by extending the entailment relation on it. This construction is made in such a way that the resulting structure is a complete *algebraic* lattice, which ensures the effectiveness of the extended entailment relation. In this paper we abstract from this construction, and we just consider the resulting structure.

Definition 4.1 (**Constraint system**) *A constraint system is a complete algebraic lattice* $\langle \mathcal{C}, \leq, \wedge, \mathit{true}, \mathit{false} \rangle$ *where* $\wedge$ *is the lub operation, and true, false are the least and the greatest elements of $\mathcal{C}$, respectively.*

In this framework, $\leq$ *is the reverse of the entailment relation* $\vdash$. *Formally:* $\forall c, d \in \mathcal{C}.\ c \vdash d \Leftrightarrow d \leq c$. *Inconsistency is modeled by false: two elements* c, c' *are consistent iff* $c \wedge c' \neq \mathit{false}$.

In order to model local variables we need the existential quantifier. An elegant way to formalize the existential quantification in the constraint framework is through the theory of cylindric algebras of Henkin et al. (1971). This leads to the concept of *cylindric constraint system*.

In the following, *Var* is a (denumerable) set of variables with typical elements $x, y, z, \ldots$.

Definition 4.2 (**Cylindric constraint system**) *Let $\langle \mathcal{C}, \leq, \wedge, true, false \rangle$ be a constraint system. Assume that for each $x \in Var$ a function $\exists_x : \mathcal{C} \to \mathcal{C}$ is defined such that for any $c, d \in \mathcal{C}$:*

(i) $\exists_x(c) \leq c$,
(ii) *if* $d \leq c$ *then* $\exists_x(d) \leq \exists_x(c)$,
(iii) $\exists_x(c \wedge \exists_x(d)) = \exists_x(c) \wedge \exists_x(d)$,
(iv) $\exists_x(\exists_y(c)) = \exists_y(\exists_x(c))$.

Then $\mathbf{C} = \langle \mathcal{C}, \leq, \wedge, true, false, Var, \exists \rangle$ *is a cylindric constraint system.*

In the following $\exists_x(c)$ is denoted by $\exists_x c$ with the convention that, in case of ambiguity, the scope of $\exists_x$ is limited to the first constraint subexpression. (So, for instance, $\exists_x c \wedge d$ stands for $\exists_x(c) \wedge d$.) Furthermore, for $\bar{x} = x_1, \ldots, x_n$ the notation $\exists_{\bar{x}}(A)$ stands for $\exists_{x_1}(\ldots \exists_{x_n}(A)\ldots)$.

In order to model parameter passing, it will be useful to consider the so-called *diagonal formulas* (Henkin et al. 1971). We assume that, for x, y ranging in Var, $\mathcal{A}$ contains the *diagonal elements* d_{xy} which satisfy the following properties.

(i) $d_{xx} = true$,
(ii) if $z \neq x$ and $z \neq y$ then $d_{xy} = \exists_z(d_{xz} \wedge d_{zy})$,
(iii) if $x \neq y$ then $c \leq d_{xy} \wedge \exists_x(c \wedge d_{xy})$.

Note that if $\mathbf{C}$ models the equality theory, then the elements d_{xy} can be thought of as the formulas $x = y$. In the following, given $\bar{x} = x_1, \ldots, x_n$ and $\bar{y} = y_1, \ldots, y_n$, we use the notation $d_{\bar{x}\bar{y}}$ to represent $d_{x_1 y_1} \wedge \cdots \wedge d_{x_n y_n}$.

4.2 The validation rule

In ALPS clauses and goals have the form

$$A \leftarrow \bar{c} \,|\, \bar{B},$$
$$\leftarrow \bar{c}, \bar{A},$$

where $\bar{B}$ can contain constraints. Note that the guard part of a clause contains only constraints.

The transition system for ALPS is obtained by replacing Rule **C2** of Table 5 by Rules **C2'** and **C2''** described in Table 9. Of course, the second component of a configurations is now a constraint.

Note that **C2'** corresponds to the P-PROLOG rule. **C2''** is called *validation rule*, and it extends the GHC mechanism.

The ALPS paradigm is more flexible than P-PROLOG and GHC. Like GHC, it allows the definition of the merge process. Furthermore, thanks to the P-PROLOG component, it allows the same program to be used with different functionalities.

C2'	$\langle \leftarrow A; c \rangle \longrightarrow \langle \bar{B}; c \wedge (A = A') \wedge c' \rangle$	there exists only one clause $A' \leftarrow c' \mid \bar{B} \in P_{ren}$ s.t. $c \wedge (A = A') \wedge c' \neq \textit{false}$
C2''	$\langle \leftarrow A; c \rangle \longrightarrow \langle \bar{B}; c \wedge (A = A') \wedge c' \rangle$	there exists $A' \leftarrow c' \mid \bar{B} \in P_{ren}$ s.t. $\exists_{\bar{x}_l}(c' \wedge (A = A')) \leq c$

Table 9 The computation rules of ALPS. In **C2''**, $\bar{x}_l$ represent the local variables of the selected clause

Example 4.3 Consider the following *append* program in ALPS:

$$append([\,],l,l) \leftarrow |,$$
$$append([x \mid l_1], l_2, [x \mid l_3]) \leftarrow | \; append(l_1, l_2, l_3).$$

This program can be activated by any call of *append* which has two arguments instantiated, and it will compute the third.

In GHC it is not possible to write an *append* program having the feature above described: the kind of argument (input or output) is fixed by the program.

5 Concurrent constraint programming

Concurrent constraint programming was proposed by Saraswat (1989), Saraswat and Rinard (1990), Saraswat et al. (1991). We follow here the definition given in (Saraswat and Rinard 1990), which is more general than the later formulations, because it covers the case of atomic unification.

In this version, concurrent constraint programming can be regarded as a generalization of most of the concurrent logic languages, including the "more powerful" ones like Concurrent PROLOG.

5.1 The language

Given a set of variables *Var* with typical elements $x, y, \ldots$, we assume fixed a cylindric constraint system $\mathbf{C} = \langle \mathcal{C}, \leq, \wedge, \textit{true}, \textit{false}, \textit{Var}, \exists \rangle$. The description of the language and the semantical definitions are parametric with respect to the constraint system.

Processes are described by the following grammar:

$$\begin{array}{lll} Processes & P ::= D.A \\ Declarations & D ::= \epsilon \mid p(\bar{x}) \text{ :- } A \mid D, D \\ Guards & g ::= \text{tell}(c) \mid \text{ask}(c) \\ Agents & A ::= \text{Stop} \mid \sum_{i=1}^{n} g_i \to A_i \mid A \parallel A \mid \exists_x A \mid p(\bar{x}) \end{array}$$

The agent **Stop** represents successful termination. The basic actions are given by the **ask**(c) and **tell**(c) constructs, where c is a *finite constraint*, i.e. an algebraic element of $\mathcal{C}$. These actions work on a common *store* which ranges over $\mathcal{C}$. The action **ask**(c) is a test on the current store and its execution does not modify the store. We say that **ask**(c) is *enabled* in d iff $c \leq d$. An action **ask**(c) *fails* in d if $c \wedge d = \textit{false}$. The action **tell**(c) is enabled in the current store d if $c \wedge d \neq \textit{false}$ and its execution sets the store to $c \wedge d$, otherwise it fails. The *guarded choice* agent $\sum_{i=1}^{n} g_i \to A_i$ selects nondeterministically one g_i which is enabled, and then behaves like A_i. If no guards are enabled, then it *fails* in case all guards fail, otherwise it *suspends*, waiting for other (parallel) agents to add information to the store. Parallel composition is represented by $\parallel$. The situation in which all components of a system of parallel agents suspend is called *global suspension* or *deadlock*. The agent $\exists_x A$ behaves like A, with x considered *local* to A. Finally, the agent $p(\bar{x})$ is a procedure call, where p is the name of the procedure and $\bar{x}$ is the list of the actual parameters. The meaning of $p(\bar{x})$ is given by a procedure declaration of the form $p(\bar{y})$:- A, where $\bar{y}$ is the list of the formal parameters. We assume that in a process D.A there is one and only one procedure declaration for every procedure name occurring in A; this is not a restriction with respect to concurrent logic languages, as the presence of alternative clauses can be simulated by the choice construct. In the following, we omit the declaration part when it is empty. Furthermore, we use **tell**(c) as syntactic sugar for **tell**(c) → **Stop**.

5.2 Operational semantics

The operational model of ccp, informally introduced above, is described by a transition system $T = (\textit{Conf}, \longrightarrow)$. The configurations (in) *Conf* are pairs consisting of a process or a guard or a *termination mode*, and a constraint representing the store. Tables 10 to 12 describe the rules of T with respect to a given set of declarations D. Rules **R2**, ..., **R6** describe the execution of the basic actions. The guarded choice operator models global non-determinism (Rules **R7**, ..., **R9**), in the sense that it depends on the current store whether or not a guard is enabled, and the current store is subject to modifications by the external environment as described by Rule **R10**. With respect to Rules **R8** and **R9** it is to be understood that in case $n = 1$ the premise of both rules reduces to $\langle g_1, c \rangle \longrightarrow \langle \tau, c \rangle$, where

R1	$\langle \text{Stop}, c \rangle \longrightarrow \langle ss, c \rangle$		
R2	$\langle \text{tell}(c), d \rangle \longrightarrow \langle ss, c \wedge d \rangle$		$c \wedge d \neq \textit{false}$
R3	$\langle \text{tell}(c), d \rangle \longrightarrow \langle \textit{ff}, c \rangle$		$c \wedge d = \textit{false}$
R4	$\langle \text{ask}(c), d \rangle \longrightarrow \langle ss, d \rangle$		$c \leq d$
R5	$\langle \text{ask}(c), d \rangle \longrightarrow \langle \textit{dd}, d \rangle$		$c \not\leq d$ and $c \wedge d \neq \textit{false}$
R6	$\langle \text{ask}(c), d \rangle \longrightarrow \langle \textit{ff}, d \rangle$		$c \wedge d = \textit{false}$

Table 10 The transition system T: basic actions

$\tau = \textit{ff}, \textit{dd}$, respectively. Rule **R10** describes parallelism as interleaving.

In order to describe locality (Rules **R13** and **R14**) the syntax has been extended by an agent $\exists_x^d A$ in which x is local to A and d is the store that has been produced locally on x. Initially the local store is empty, i.e. $\exists_x A = \exists_x^{\textit{true}} A$. Intuitively, a process $\exists_x A$ can make a step in the store c if the process A can make a step in the store $\exists_x c$, since at local level A does not see the global information on x. If the local computation has already produced some constraint d on x, that will be considered as well, i.e. the store seen locally is $d \wedge \exists_x c$. The constraint d' produced by this step is added to the local store, and $\exists_x d'$ (i.e. d' with the information on x removed) is added to the global store.

The execution of a procedure call is modeled by **R15**. $\Delta_{\bar{y}}^{\bar{x}}$ stands for $\exists_{\bar{\alpha}}^{d_{\bar{x}\bar{\alpha}}} \exists_{\bar{y}}^{d_{\bar{\alpha}\bar{y}}}$ and it is used to establish the link between the formal parameters $\bar{y}$ and the actual parameters $\bar{x}$. The variables $\bar{\alpha}$ are introduced in order to avoid problems related to names clash between $\bar{x}$ and $\bar{y}$. They are assumed to be all distinct and not to occur in the program.

In the following we assume the set of declarations to be fixed.

The notion of observables we consider here is the collection of the final results of the computations, together with the termination modes. For the sake of simplicity we consider only finite computations.

Concurrent logic and concurrent constraint programming

$$\text{R7} \quad \frac{\langle g_i, c \rangle \longrightarrow \langle ss, d \rangle}{\langle \sum_{i=1}^{n} g_i \to A_i, c \rangle \longrightarrow \langle A_i, d \rangle}$$

$$\text{R8} \quad \frac{\langle g_n, c \rangle \longrightarrow \langle \mathit{ff}, c \rangle \quad \langle \sum_{i=1}^{n-1} g_i \to A_i, c \rangle \longrightarrow \langle \tau, c \rangle}{\langle \sum_{i=1}^{n} g_i \to A_i, c \rangle \longrightarrow \langle \tau, c \rangle} \quad \tau \in \{\mathit{ff}, \mathit{dd}\}$$

$$\text{R9} \quad \frac{\langle g_n, c \rangle \longrightarrow \langle \mathit{dd}, c \rangle \quad \langle \sum_{i=1}^{n-1} g_i \to A_i, c \rangle \longrightarrow \langle \tau, c \rangle}{\langle \sum_{i=1}^{n} g_i \to A_i, c \rangle \longrightarrow \langle \mathit{dd}, c \rangle} \quad \tau \in \{\mathit{ff}, \mathit{dd}\}$$

Table 11 The transition system T: guarded choice

Definition 5.1 *The observables are given by the function*

$$\mathcal{O}(A) = \{ \langle c, \tau \rangle \mid \langle A, \mathit{true} \rangle \longrightarrow^+ \langle \tau, c \rangle \}$$

where $\longrightarrow^+$ *denotes the transitive closure of* $\longrightarrow$. *We don't need to consider here the reflexive closure because every agent makes at least one step before reaching a final configuration.*

Note that from this definition we can derive the input output behavior of a process. In fact, the results of A in an initial store $c \neq \mathit{true}, \mathit{false}$ coincide with the results of the process $\mathbf{tell}(c) \parallel A$ in the initial store true.

5.3 Embedding concurrent logic programming

Concurrent constraint programming subsumes the main flat concurrent logic languages. We show here how Flat GHC and FCP (a version of Flat CP, proposed by Kliger *et al.* (1988)) can be translated into ccp.

We will use the following notation: if $\bar{t} = t_1, \ldots, t_n$ and $\bar{u} = u_1, \ldots, u_n$ then $\bar{t} = \bar{u}$ stands for $t_1 = u_1 \wedge \cdots \wedge t_n = u_n$.

5.3.1 Embedding Flat GHC

A Flat GHC goal $\leftarrow A_1, \ldots, A_n$ corresponds to the ccp agent $B_1 \parallel \cdots \parallel B_n$; where, if A_i is the atom $p(\bar{t})$, then B_i is the agent $\exists_{\bar{x}}(p(\bar{x}) \parallel \mathbf{tell}(\bar{x} = \bar{t}))$ with $\bar{x}$ fresh distinct variables.

R10 $\dfrac{\langle A,c\rangle \longrightarrow \langle A',d\rangle}{\langle A \parallel B, c\rangle \longrightarrow \langle A' \parallel B, d\rangle}$

$\langle B \parallel A, c\rangle \longrightarrow \langle B \parallel A', d\rangle$

R11 $\dfrac{\langle A,c\rangle \longrightarrow \langle \mathit{ff},c\rangle}{\langle A \parallel B, c\rangle \longrightarrow \langle \mathit{ff}, c\rangle}$

$\langle B \parallel A, c\rangle \longrightarrow \langle \mathit{ff}, c\rangle$

R12 $\dfrac{\langle A,c\rangle \longrightarrow \langle \mathit{dd},c\rangle \ \ \langle B,c\rangle \longrightarrow \langle \mathit{dd},c\rangle}{\langle A \parallel B, c\rangle \longrightarrow \langle \mathit{dd}, c\rangle}$

R13 $\dfrac{\langle A, d \wedge \exists_x c\rangle \longrightarrow \langle B, d'\rangle}{\langle \exists_x^d A, c\rangle \longrightarrow \langle \exists_x^{d'} B, c \wedge \exists_x d'\rangle}$

R14 $\dfrac{\langle A, d \wedge \exists_x c\rangle \longrightarrow \langle \tau, d'\rangle}{\langle \exists_x^d A, c\rangle \longrightarrow \langle \tau, c \wedge \exists_x d'\rangle}$

R15 $\langle p(\bar{x}), c\rangle \longrightarrow \langle \Delta_{\bar{y}}^{\bar{x}} A, c\rangle$ $\qquad\qquad p(\bar{y})\text{ :- } A$ is in D

Table 12 The transition system T: parallelism, hiding and procedure call

Concerning predicate definitions, let p be defined by the following Flat GHC clauses:

$$p(\bar{t}_1) \ \leftarrow \ \bar{u}_1 = \bar{u}'_1 \mid \bar{A}_1, (\bar{v}_1 = \bar{v}'_1), \bar{A}'_1,$$
$$\vdots$$
$$p(\bar{t}_n) \ \leftarrow \ \bar{u}_n = \bar{u}'_n \mid \bar{A}_n, (\bar{v}_n = \bar{v}'_n), \bar{A}'_n.$$

Then p can be equivalently defined by the following ccp program:

$$p(\bar{x}) \;\; :\text{-} \;\; \mathbf{ask}(\exists_{\bar{y}}(\bar{x} = \bar{t_1} \wedge \bar{u}_1 = \bar{u}'_1)) \rightarrow E_1$$

$$+$$

$$\vdots$$

$$+$$

$$\mathbf{ask}(\exists_{\bar{y}}(\bar{x} = \bar{t_n} \wedge \bar{u}_n = \bar{u}'_n)) \rightarrow E_n,$$

where $\bar{x}$ are fresh distinct variables and $\bar{y}$ are all the other variables occurring in the original definition. The E_i's are the agents

$$E_i \;\; = \;\; \exists_{\bar{y}}(\mathbf{tell}(\bar{x} = \bar{t_i} \wedge \bar{u}_i = \bar{u}'_i) \parallel \bar{B}_i \parallel \mathbf{tell}(\bar{v}_i = \bar{v}'_i) \parallel \bar{B}'_i\,),$$

where the $\bar{B}_i$'s and the $\bar{B}'_i$'s are obtained from the goals $\leftarrow \bar{A}_i$ and $\leftarrow \bar{A}'_i$ respectively as specified above.

Note that the ask-rule in ccp is more flexible than in ALPS: the quantification of the local variables is not done automatically by the evaluation of the ask, therefore we must explicitly put the existential quantifier into the guards, and tell the constraints afterwards in order to establish the links with the local variables.

Example 5.2 Angelic merge can be defined in ccp as follows.

$$merge(l_1, l_2, l_3) \;\; :\text{-} \;\; \mathbf{ask}(\exists_{x,l'_1}\; l_1 = [x|l'_1]) \rightarrow \exists_{x,l'_1,l'_3}(\mathbf{tell}(l_1 = [x|l'_1]) \parallel$$

$$\mathbf{tell}(l_3 = [x|l'_3]) \parallel$$

$$merge(l'_1, l_2, l'_3)\,)$$

$$+$$

$$\mathbf{ask}(\exists_{y,l'_2}\; l_2 = [y|l'_2]) \rightarrow \exists_{y,l'_2,l'_3}(\mathbf{tell}(l_2 = [y|l'_2]) \parallel$$

$$\mathbf{tell}(l_3 = [y|l'_3]) \parallel$$

$$merge(l_1, l'_2, l'_3)\,)$$

$$+$$

$$\mathbf{ask}(l_1 = [\,]) \rightarrow \mathbf{tell}(l_3 = l_2)$$

$$+$$

$$\mathbf{ask}(l_2 = [\,]) \rightarrow \mathbf{tell}(l_3 = l_1)$$

5.3.2 Embedding FCP

FCP is a variant of Flat Concurrent PROLOG in which input channels are specified statically. A typical FCP clause is of the form

$$p(\bar{x}) \leftarrow \bar{g}_1 : \bar{g}_2 \mid \bar{A},$$

where $\bar{g}_1$ is the input guard, i.e. a sequence of equalities which must be unifiable without introducing new global bindings, and $\bar{g}_2$ is the output guard, namely a sequence of equalities whose unifier must be compatible with the current bindings. The evaluation of $\bar{g}_2$ corresponds to the atomic unification, i.e. it contains the consistency check and introduces new bindings before commitment.

The translation of a FCP goal into ccp is defined as in the Flat GHC case. Concerning the programs, we consider here, for the sake of simplicity, only the case in which each clause contains either $\bar{g}_1$ or $\bar{g}_2$, but not both. The interested reader is invited to extend the translation to the full case.

Let p be defined by the following FCP clauses:

$$p(\bar{x}) \leftarrow \bar{u}_1 = \bar{u}_1' :\mid \bar{A}_1, \bar{v}_1 = \bar{v}_1', \bar{A}_1',$$
$$\vdots$$
$$p(\bar{x}) \leftarrow : \bar{u}_n = \bar{u}_n' \mid \bar{A}_n, \bar{v}_n = \bar{v}_n', \bar{A}_n'.$$

Then p can be equivalently defined by the following ccp program:

$$\begin{aligned} p(\bar{x}) \; :\text{-} \; & \mathbf{ask}(\exists_{\bar{y}} \bar{u}_1 = \bar{u}_1') \to E_1 \\ & + \\ & \vdots \\ & + \\ & \mathbf{tell}(\exists_{\bar{y}} \bar{u}_n = \bar{u}_n') \to E_n \end{aligned} ,$$

where the E_i's are the agents

$$E_i \;=\; \exists_{\bar{y}}(\mathbf{tell}(\bar{u}_i = \bar{u}_i') \parallel \bar{B}_i \parallel \mathbf{tell}(\bar{v}_i = \bar{v}_i') \parallel \bar{B}_i'),$$

where the y_i's, the $\bar{B}_i$'s and the $\bar{B}_i''$'s are defined as in the Flat GHC case.

5.4 Denotational semantics

In this section we address the problem of a compositional semantics for concurrent constraint programming, correct with respect to the observables $\mathcal{O}$ defined in the previous section. Compositionality is considered one of the most desirable characteristics of a formal semantics, since it provides a foundation of program verification and modular design.

First we observe that $\mathcal{O}$ is not compositional:

Example 5.3 Let $A_1 = \mathbf{ask}(c) \to \mathbf{Stop}$ and $A_2 = \mathbf{ask}(d) \to \mathbf{Stop}$. We have that $\mathcal{O}(A_1) = \mathcal{O}(A_2) = \{\langle true, dd \rangle\}$. However, given the agent $B = \mathbf{tell}(c) \to \mathbf{Stop}$, and assuming $c \not\leq d$, we have that $\mathcal{O}(A_1 \parallel B) \neq \mathcal{O}(A_2 \parallel B)$, since $\langle true, c \rangle \in \mathcal{O}(A_1 \parallel B) \setminus \mathcal{O}(A_2 \parallel B)$.

Therefore we have to introduce some additional information which, for example, allows us to distinguish between the above two agents A_1 and A_2. A first natural attempt seems to add information about the *input output* behavior: define $\mathcal{IO}$ as follows:

$$\mathcal{IO}(A) = \{\langle c, d, \tau \rangle \mid \langle A, c \rangle \longrightarrow^{+} \langle \tau, d \rangle\}.$$

It is easy to check that indeed the semantics $\mathcal{IO}$ distinguishes between the agents A_1 and A_2. However, still the semantics $\mathcal{IO}$ does not provide the necessary information about the *branching structure* of agents. Consider the following counterexample (by Falaschi et al. (1993)) to the compositionality of $\mathcal{IO}$:

Example 5.4 Let $a, b, c \in \mathcal{C}$ be constraints ordered by the relation $a \leq b \leq c$ and consider the agents A_1 and A_2:

$$A_1 = \quad \mathbf{ask}(true) \to \mathbf{tell}(a)$$
$$+$$
$$\mathbf{ask}(b) \to \mathbf{tell}(c)$$

$$A_2 = \quad \mathbf{ask}(true) \to \mathbf{tell}(a)$$
$$+$$
$$\mathbf{ask}(b) \to \mathbf{tell}(c)$$
$$+$$
$$\mathbf{ask}(true) \to \mathbf{tell}(a) \to \mathbf{ask}(b) \to \mathbf{tell}(c)$$

It is easy to check that $\mathcal{IO}(A_1) = \mathcal{IO}(A_2)$. However, given the agent

$$B = \mathbf{ask}(a) \to \mathbf{tell}(b)$$

we have $\mathcal{IO}(A_1 \parallel B) \neq \mathcal{IO}(A_2 \parallel B)$ since $\langle true, c \rangle \in \mathcal{IO}(A_2 \parallel B) \setminus \mathcal{IO}(A_1 \parallel B)$.

The problem in the example above is that, when considering the input output behavior, the (agent corresponding to the) third branch of the agent A_2 is equivalent to the union of the first two branches. Indeed, the control structure disappears and the third branch is represented by the two pairs $\langle true, a \rangle$ and $\langle b, c \rangle$.

5.4.1 Reactive sequences

De Boer and Palamidessi (1991a) have shown that the necessary branching information is encoded by a generalization of the input output behavior in terms of *sequences* of pairs of assume tell constraints. Such sequences are called reactive sequences and describe the interaction between a process and its environment. Reactive sequences will be represented by finite sequences of labeled constraints $s, s', \ldots$. Updates to the common store by the environment will be indicated by the label a (*assume*), while updates by the process itself will be labeled by t (*tell*). We use ℓ to range over $\{a, t\}$.

The *information content* of a reactive sequence s is the conjunction of all its constraints, and represents the store as determined by the sequence of events encoded by the sequence. The information content of s will be indicated by $con(s)$. Formally:

(i) $con(\lambda) = true$ (λ is the empty sequence), and
(ii) $con(c^\ell \cdot s) = c \wedge con(s)$.

To define the compositional semantics based on reactive sequences we use a transition system T' (see Tables 13-16). The configurations are pairs $\langle A, s \rangle$, where s is a reactive sequence such that $con(s)$ is consistent. The difference with the transition system T consists mainly in Rule **R16'**, which models the interaction with the environment. Note that a process A is not immediately affected by actions made by the environment (only its future behavior will depend on them). An arbitrary constraint can be added (by the environment) consistently to the store without changing the state of A. In other words, A can make an *arbitrary assumption* (provided it is consistent) about the store. The other rules correspond to the rules of T. In the modified rules for the guarded choice the operation $\exists_x$ applied to a sequence s denotes the sequence obtained from s by applying $\exists_x$ to the constraints of s. Furthermore, $s_{\restriction \ell}$ denotes the subsequence of s consisting of the constraints labeled by ℓ.

The correspondence between T and T' is expressed by the following lemma.

Lemma 5.5 *(de Boer and Palamidessi 1991a) Rules* **R1'-R15'** *of T' mimic Rules* **R1-R15** *of T, in the sense that if*

$$\langle A, s \rangle \longrightarrow \langle A', s' \rangle$$

is an **Ri'** *transition step in T', then*

$$\langle A, con(s) \rangle \longrightarrow \langle A', con(s') \rangle$$

is an **Ri** *transition step in T.*

We show now that we can obtain a compositional semantics $\mathcal{M}$ by col-

R1'	$\langle \text{Stop}, s \rangle \longrightarrow \langle ss, s \rangle$		
R2'	$\langle \text{tell}(c), s \rangle \longrightarrow \langle ss, s \cdot c^t \rangle$	$con(s) \wedge c \neq \mathit{false}$	
R3'	$\langle \text{tell}(c), s \rangle \longrightarrow \langle \mathit{ff}, s \rangle$	$con(s) \wedge c = \mathit{false}$	
R4'	$\langle \text{ask}(c), s \rangle \longrightarrow \langle ss, s \rangle$	$c \leq con(s)$	
R5'	$\langle \text{ask}(c), s \rangle \longrightarrow \langle dd, s \rangle$	$c \not\leq con(s)$ and $con(s) \wedge c \neq \mathit{false}$	
R6'	$\langle \text{ask}(c), s \rangle \longrightarrow \langle \mathit{ff}, s \rangle$	$con(s) \wedge c = \mathit{false}$	

Table 13 The transition system T': basic actions

lecting, for every process, the sets of reactive sequences generated by the transition system T'. We need of course to consider all the completed sequences, and we will signal the way the process has terminated by postfixing a termination mode $\tau \in \{ss, \mathit{ff}, dd\}$. However, for a compositional treatment of failure in the presence of infinite processes, we need also to collect the sequences corresponding to unfinished computations. Such sequences will be postfixed by the symbol $\bot$.

Example 5.6 Consider the process $p(x)$, with $p(x)$:- $\text{tell}(x = f(z)) \rightarrow p(z)$. Since there are no terminating computations of $p(x)$, in the absence of a representation of unfinished computations, its semantics will be empty. However, when $p(x)$ is put in parallel with $\text{tell}(\mathit{false})$, we get as observable results all the pairs $\langle \exists_z x = f^n(z), \mathit{ff} \rangle$, where $f^0(z) = f(z)$ and $f^{n+1}(z) = f(f^n(z))$.

So, despite our interest in finite computations only, we need to represent the finite approximations of infinite computations for a proper treatment of failure.

$$
\textbf{R7}' \quad \frac{\langle g_i, s\rangle \longrightarrow \langle ss, s'\rangle}{\langle \sum_{i=1}^{n} g_i \to A_i, s\rangle \longrightarrow \langle A_i, s'\rangle}
$$

$$
\textbf{R8}' \quad \frac{\langle g_n, s\rangle \longrightarrow \langle \mathit{ff}, s\rangle \quad \langle \sum_{i=1}^{n-1} g_i \to A_i, s\rangle \longrightarrow \langle \tau, s\rangle}{\langle \sum_{i=1}^{n} g_i \to A_i, s\rangle \longrightarrow \langle \tau, s\rangle} \quad \tau \in \{\mathit{ff}, \mathit{dd}\}
$$

$$
\textbf{R9}' \quad \frac{\langle g_n, s\rangle \longrightarrow \langle \mathit{dd}, s\rangle \quad \langle \sum_{i=1}^{n-1} g_i \to A_i, s\rangle \longrightarrow \langle \tau, s\rangle}{\langle \sum_{i=1}^{n} g_i \to A_i, s\rangle \longrightarrow \langle \mathit{dd}, s\rangle} \quad \tau \in \{\mathit{ff}, \mathit{dd}\}
$$

Table 14 The transition system T': guarded choice

Definition 5.7 *The semantics $\mathcal{M}$ is defined as follows:*

$$
\begin{aligned}
\mathcal{M}[\![A]\!] = \quad & \{s \cdot ss \quad \mid \langle A, \mathit{true}\rangle \longrightarrow^{+} \langle ss, s\rangle\} \\
\cup \quad & \{s \cdot \mathit{ff} \quad \mid \langle A, \mathit{true}\rangle \longrightarrow^{+} \langle \mathit{ff}, s\rangle\} \\
\cup \quad & \{s \cdot \mathit{dd} \quad \mid \langle A, \mathit{true}\rangle \longrightarrow^{+} \langle \mathit{dd}, s\rangle\} \\
\cup \quad & \{s \cdot \bot \quad \mid \langle A, \mathit{true}\rangle \longrightarrow^{+} \langle A', s\rangle\}.
\end{aligned}
$$

The observables can be retrieved from $\mathcal{M}$ by first selecting those sequences entirely composed of tell constraints and ended by ss, ff, or dd. This amounts to requiring that the process has run in isolation and that the computation has reached a final state. Then we abstract from the particular order in which the constraints have been produced, and close under logical equivalence. This procedure is described by the operator *result*:

$$
\mathit{result}(S) = \{\langle \mathit{con}(s), \tau\rangle \mid s \cdot \tau \in S \ \wedge \ s = c_1^t \cdots c_n^t \ \wedge \ \tau \in \{ss, \mathit{ff}, \mathit{dd}\}\}.
$$

From Lemma 5.5 we obtain the correctness of the model with respect to our notion of observables:

Concurrent logic and concurrent constraint programming

$$
\textbf{R10}' \quad \frac{\langle A, s\rangle \longrightarrow \langle A', s'\rangle}{\begin{array}{l}\langle A \parallel B, s\rangle \longrightarrow \langle A' \parallel B, s'\rangle \\ \langle B \parallel A, s\rangle \longrightarrow \langle B \parallel A', s'\rangle\end{array}}
$$

$$
\textbf{R11}' \quad \frac{\langle A, s\rangle \longrightarrow \langle \mathit{ff}, s\rangle}{\begin{array}{l}\langle A \parallel B, s\rangle \longrightarrow \langle \mathit{ff}, s\rangle \\ \langle B \parallel A, s\rangle \longrightarrow \langle \mathit{ff}, s\rangle\end{array}}
$$

$$
\textbf{R12}' \quad \frac{\langle A, s\rangle \longrightarrow \langle \mathit{dd}, s\rangle \quad \langle B, s\rangle \longrightarrow \langle \mathit{dd}, s\rangle}{\langle A \parallel B, s\rangle \longrightarrow \langle \mathit{dd}, s\rangle}
$$

$$
\textbf{R13}' \quad \frac{\langle A, (\exists_x s)\cdot d^t\rangle \longrightarrow \langle B, s'\rangle}{\langle \exists_x^d A, s\rangle \longrightarrow \langle \exists_x^{con(s')} B, s\cdot(\exists_x con(s'))^t\rangle} \quad \text{where } s_{\restriction a} = s'_{\restriction a}
$$

$$
\textbf{R14}' \quad \frac{\langle A, (\exists_x s)\cdot d^t\rangle \longrightarrow \langle \tau, s'\rangle}{\langle \exists_x^d A, s\rangle \longrightarrow \langle \tau, s\cdot(\exists_x con(s'))^t\rangle}
$$

$$
\textbf{R15}' \quad \langle p(\bar{x}), s\rangle \longrightarrow \langle \Delta_{\bar{y}}^{\bar{x}} A, s\rangle \qquad p(\bar{y}) \text{ :- } A \text{ is in } D
$$

Table 15 The transition system T': parallelism, hiding and procedure call

$$
\textbf{R16}' \quad \langle A, s\rangle \longrightarrow \langle A, s\cdot c^a\rangle \qquad c \wedge con(s) \neq \mathit{false}
$$

Table 16 The transition system T': assumption rule

Theorem 5.8 (Correctness of $\mathcal{M}$, (de Boer and Palamidessi 1991a))

$$\mathcal{O}(A) = \mathit{result}(\mathcal{M}[\![A]\!]).$$

The semantics $\mathcal{M}$ is denotational, i.e. compositional with respect to all the operators of the language. Here we show only the semantic counterparts of the parallel and the hiding operators, the interested reader can find the definition of the other operators in (de Boer and Palamidessi 1991a).

We first consider the parallel operator. A reactive sequence represents a particular sequence of reactions (encoded by the tell constraints) of a process to a particular sequence of actions (encoded by the assume constraints) performed by a hypothetic environment. The model $\mathcal{M}$ gives the set of all these possible sequences. Therefore it encodes already all the outcomes of all possible interactions with a parallel process. To obtain such outcomes the idea is to combine each sequence of the first process with each sequence of the second which is compatible, in the sense that at every point the action of the first corresponds to an assumption by the second and vice versa. Also, we allow the presence of corresponding assumptions on both sides, since they encode possible future interactions with another process. "To correspond" here means to be relative to the same constraint; in fact our model is based on the assumption of a global store, hence at every step all processes must see the same store. In other words, the parallel operator "unzips" the sequences, i.e. it verifies that the assumptions made by a process are validated by the other. The compatible sequences are then combined according to the principle that if one of the two constraints is a tell (i.e. it is produced by at least one of the two processes) then the result is a tell, otherwise it is an assume.

Concerning the termination modes, the result is success if both processes successfully terminate. It is deadlock if one deadlocks and the other either deadlocks or terminates, and it is failure if at least one fails. In fact the detection of a failure propagates through the whole system.

The parallel operator described above is similar to an interleaving operator. The difference is that it applies to sequences containing already all the information concerning the way in which processes interleave (the assumptions specify "where" and "what"). Such an operator was introduced by Gerth *et al.* (1988), following ideas of Ever-Hadani (1987).

Formally we have the following definition.

Definition 5.9 *We define the (partial) parallel operator $\parallel$ on reactive sequences (together with a termination mode) as follows:*

(i) $s_1 \cdot T_1 \parallel s_2 \cdot T_2 = s_2 \cdot T_1 \parallel s_1 \cdot T_2$
(ii) $c^a \cdot s_1 \cdot T_1 \parallel c^\ell \cdot s_2 \cdot T_2 = c^\ell \cdot (s_1 \cdot T_1 \parallel s_2 \cdot T_2)$
(iii) $\tau \parallel ss = \tau$
(iv) $\tau \parallel f\!\!f = f\!\!f$
(v) $dd \parallel dd = dd$
(vi) $\perp \parallel \perp = \perp$.

Since arbitrary assumptions can always be made, it is sufficient to consider the cases listed above, and to leave $\parallel$ undefined on the sequences of different kind. The extension of the parallel operator to sets is defined in the obvious

way:
$$S_1 \tilde{\|} S_2 = \{s_1 \cdot \tau_1 \tilde{\|} s_2 \cdot \tau_2 \mid s_1 \cdot \tau_1 \in S_1 \wedge s_2 \cdot \tau_2 \in S_2\}.$$

Next we introduce the semantic hiding operator. Intuitively, the hiding operator, applied to a set S of reactive sequences denoting a process, should filter out all those sequences which make assumptions on x, since the global information on x has no influence at the local level. Conversely, the information on x contained in the tell constraints will be removed, since the local information produced on x will not be visible at the global level. In conclusion, for every sequence $s \in S$ the hiding operator will deliver all those sequences s' which have at every point the the same assume constraints, once the information on x is removed from the constraint in s', and the same tell constraints, after the information on x is removed from the constraint in s.

Definition 5.10 *Given a set of reactive sequences (together with a termination mode) we define:*

$$\exists_x(S) = \{s \mid \text{ there exists } s' \in S \text{ s.t. } \exists_x(s\!\restriction_a) = s'\!\restriction_a \text{ and } \exists_x(s'\!\restriction_t) = s\!\restriction_t\}$$

where $\exists_x s$ denotes the pointwise extension to (labeled) sequences of constraints of $\exists_x$.

By a straightforward case analysis of the transition system T' we have the following theorem:

Theorem 5.11 (**Compositionality of** $\mathcal{M}$ *(de Boer and Palamidessi 1991a)*)

(i) $\mathcal{M}[\![A \parallel B]\!] = \mathcal{M}[\![A]\!] \parallel \mathcal{M}[\![B]\!]$,
(ii) $\mathcal{M}[\![\exists_x A]\!] = \exists_x(\mathcal{M}[\![A]\!])$.

5.4.2 Full abstraction

The compositional semantics $\mathcal{M}$ is not fully abstract with respect to the observables. Namely, it distinguishes processes which are observationally equivalent in any context. The problem is that sequences contain redundant information about the order and the granularity in which constraints are added to the store.

Example 5.12 Let A and B be the agents shown in Figure 7 and defined as follows:

$$A = \text{tell}(c_1) \to \text{tell}(c_2)$$
$$+$$
$$\text{tell}(c_1 \wedge c_2)$$

$$B = \text{tell}(c_1) \to \text{tell}(c_2) \to \text{Stop}.$$

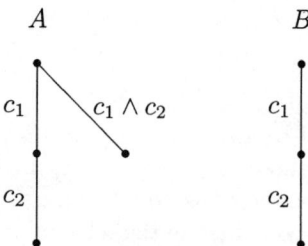

FIG. 7. Two agents observationally equivalent in every context

We have that $\mathcal{M}[\![A]\!] \neq \mathcal{M}[\![B]\!]$. Nevertheless they will produce the same final results in every context. This is because of the monotonicity of communication: we cannot define an environment that accepts both c_1 and c_2 and not $c_1 \wedge c_2$.

In general, the reaction of a process to a certain sequence of actions will only depend upon the information content of the sequence. Therefore, to obtain full abstraction we must eliminate distinctions between equivalent subsequences (equivalent in the sense of having the same information content). One way to do this is to *saturate* a set of denotations by adding, for any sequence s, all those sequences that differ only for an equivalent subsequence (**C1**). This operation, however, is not enough: since (**C1**) will introduce new possible interleaving points we need a rule (**C2**) which adds assumptions in order to make those interleaving points explicit for the parallel operator. These assumptions of course must not add any new information.

Definition 5.13 *Given a set of reactive sequences S, the saturation of S, $Sat(S)$, is the minimal set of sequences which contain S and satisfies the conditions* **C1** *and* **C2** *of Table 17.*

We then define the fully abstract semantics A as follows:

$$\mathcal{A}[\![A]\!] = Sat(\mathcal{M}[\![A]\!]).$$

In (de Boer and Palamidessi 1991a) $\mathcal{A}$ is proved to be correct, compositional and fully abstract.

5.5 Reactive sequences versus closure operators

In this section we show that if we eliminate the consistency check, and we drop the distinction between success and deadlock, then the model $\mathcal{M}$ of the previous sections can be expressed in terms of Scott's *closure operators*. On this basis we prove the isomorphism of our restricted model with the model developed by Saraswat et al. (1991).

> **C1** if $s_1 \cdot c_1^\ell \ldots c_n^\ell \cdot s_2 \cdot \tau \in S$ and $con(s_1 \cdot c_1^\ell \ldots c_n^\ell) = con(s_1 \cdot d_1^\ell \ldots d_m^\ell)$
> then $s_1 \cdot d_1^\ell \ldots d_m^\ell \cdot s_2 \cdot \tau \in S$
>
> **C2** if $s_1 \cdot s_2 \cdot \tau \in S$ and $con(s_1 \cdot s_2) = con(s_1 \cdot c \cdot s_2)$ then $s_1 \cdot c^a \cdot s_2 \cdot \tau \in S$

Table 17 The saturation conditions

The consistency check requires the application, before executing an action of the form **tell**(c), of a test on the consistency of c with respect to the current store. If c is inconsistent with the current store then we choose another possibility, and only in case there are no other possibilities then a failure is generated. When the consistency check is not supported then **tell**(c) just adds c to the store regardless of whether or not it will lead to an inconsistent situation (*false*). The two kinds of tell are called *atomic tell* and *eventual tell* respectively in (Saraswat and Rinard 1990), and they clearly reflect the distinction between atomic and eventual unification. Note that the inconsistent store is a situation of no recovery, since it is the top of the lattice, and the store can only evolve monotonically. Inconsistency is regarded as the *most undesirable* situation, the *failure* of the computation.

Concurrent constraint without consistency check can be defined by forbidding the occurrence of a tell-guard in a plus context, thus eliminating escape possibilities when a failure is detected, and by eliminating some rules in the transition system (essentially the ones concerning failure). Note that, since tell is no longer a guard, it must be introduced as basic action. Namely, **tell**(c) is now an agent.

The identification of deadlock and success in the observables means that we renounce to the possibility of distinguishing processes which have terminated and will not react to any "stimulus", from processes which are just stuck, and which will still produce information once new input is provided.

Example 5.14 Consider the two processes

$$A_1 = \text{ask}(\mathit{true}) \to \text{tell}(x = a \land y = b)$$
$$+$$
$$\text{ask}(\mathit{true}) \to \text{ask}(x = a) \to \text{tell}(x = a \land y = b)$$
$$+$$
$$\text{ask}(\mathit{true}) \to \text{Stop}$$

$$A_2 = \text{ask}(\mathit{true}) \to \text{tell}(x = a \land y = b)$$
$$+$$
$$\text{ask}(\mathit{true}) \to \text{Stop}$$

The semantics proposed by Saraswat et al. (1991) does not distinguish between A_1 and A_2, whereas the semantics presented in previous sections does. In fact, the second branch of A_1 can generate deadlock (depending upon the store) whereas this possibility is excluded in A_2. If we dropped the distinction between success and deadlock then also our model would identify A_1 and A_2.

The identification of deadlock and success is maybe questionable from the practical point of view, but it makes possible an elegant representation of the reactive sequences in terms of closure operators. The rest of this section is devoted to investigating this characterization.

The transition system for concurrent constraint without consistency check and without deadlock recognition is given in Table 18. From it we can derive the notion of observables (still denoted by $\mathcal{O}$) in the same way as before.

5.5.1 *The restricted denotational model*

To define the compositional semantics we adopt the same "pruning" (elimination of the rules for failure and deadlock) on the transition system T'. The new transition system, rT', is given in Table 19.

Based on rT' a semantics $\mathcal{M}$ is defined in the same way as before. All the results of previous section hold for $\mathcal{M}$ and $\mathcal{O}$. In particular, the compositionality of $\mathcal{M}$ and the correctness of $\mathcal{M}$ with respect to $\mathcal{O}$. Finally, we obtain the fully abstract model $\mathcal{A}$ for our sublanguage by applying the saturation operator Sat to $\mathcal{M}$. Observe that only one termination symbol remains there: ss. Thus we can drop it from the reactive sequences of $\mathcal{M}$ and $\mathcal{A}$.

rR1	$\langle \mathbf{tell}(c), d \rangle \longrightarrow \langle \mathbf{Stop}, c \wedge d \rangle$
rR2	$\langle \sum_{i=1}^{n} \mathbf{ask}(c_i) \to A_i, d \rangle \longrightarrow \langle A_j, d \rangle \quad j \in [1,n] \text{ and } c_j \leq d$
rR3	$\dfrac{\langle A, c \rangle \longrightarrow \langle A', c' \rangle}{\langle A \parallel B, c \rangle \longrightarrow \langle A' \parallel B, c' \rangle}$ $\langle B \parallel A, c \rangle \longrightarrow \langle B \parallel A', c' \rangle$
rR4	$\dfrac{\langle A, d \wedge \exists_x c \rangle \longrightarrow \langle B, d' \rangle}{\langle \exists_x^d A, c \rangle \longrightarrow \langle \exists_x^{d'} B, c \wedge \exists_x d' \rangle}$
rR5	$\langle p(\bar{x}), c \rangle \longrightarrow \langle \Delta_{\bar{y}}^{\bar{x}} A, c \rangle \qquad p(\bar{y}) :\text{-} A \text{ is in } D$

Table 18 The restricted transition system rT

5.5.2 Closure operators

A closure operator on a lattice $\langle \mathcal{C}, \leq \rangle$ is a function $\Psi : \mathcal{C} \to \mathcal{C}$ which is *idempotent*, *monotonic* and *extensive*. More formally:

(i) $\forall x \in \mathcal{C}.\ \Psi(\Psi(x)) = \Psi(x)$,
(ii) $\forall x, y \in \mathcal{C}.\ x \leq y \Rightarrow \Psi(x) \leq \Psi(y)$,
(iii) $\forall x \in \mathcal{C}.\ x \leq \Psi(x)$.

Closure operators have the following nice property: they are completely determined by their range, which coincides with the set of their fixed points. In fact, if $R \subseteq \mathcal{C}$ is the range of a closure operator Ψ, then we can retrieve Ψ as follows

$$\Psi(x) = min\{y \in R \mid x \leq y\}.$$

We can also prove that, given a set R, if every subset of R which is bounded from below admits the least element, then R is the range of a closure operator. In the following, we will identify Ψ and its range.

Closure operators are used by Saraswat *et al.* (1991) to give semantics to determinate cc programming, namely cc programming without nondeterministic choice. A priori the absence of nondeterministic choice does not imply that the language is deterministic, because the parallel operator might re-introduce nondeterminism. In the case in which the consistency

rR1'	$\langle \text{tell}(c), s \rangle \longrightarrow \langle \text{Stop}, s \cdot c^t \rangle$	
rR2'	$\langle \sum_{i=1}^{n} \text{ask}(c_i) \to A_i, s \rangle \longrightarrow \langle A_j, s \rangle$	$j \in [1, n]$ and $c_j \leq con(s)$
rR3'	$\dfrac{\langle A, s \rangle \longrightarrow \langle A', s' \rangle}{\begin{array}{c}\langle A \parallel B, s \rangle \longrightarrow \langle A' \parallel B, s' \rangle \\ \langle B \parallel A, s \rangle \longrightarrow \langle B \parallel A', s' \rangle\end{array}}$	
rR4'	$\dfrac{\langle A, d \wedge \exists_x s \cdot d^t \rangle \longrightarrow \langle B, s' \rangle}{\langle \exists_x^d A, s \rangle \longrightarrow \langle \exists_x^{con(s')} B, s \cdot (\exists_x con(s'))^t \rangle}$	
rR5'	$\langle p(\bar{x}), s \rangle \longrightarrow \langle \Delta_{\bar{y}}^{\bar{x}} A, s \rangle$	$p(\bar{y}) \coloneq A$ is in D
rR6'	$\langle A, s \rangle \longrightarrow \langle A, s \cdot c^a \rangle$	

Table 19 The transition system rT'

check is not supported, however, it is possible to prove a sort of confluence property, which causes a system to reach always the same final store from a given initial store. A determinate cc process can therefore be seen as a function from store to store. Furthermore, the way in which the process transforms the store satisfies the properties mentioned above, hence this function is actually a closure operator. The fact that closure operators can be represented as sets allows the semantic operators to be defined as operators on sets. In the case of the merge operator this is particularly convenient, in fact it just corresponds to set intersection.

Concerning the nondeterminate case, the behavior of a process can no longer be described as a *single* function from the initial store to the final store: we need at least a set of functions. This turns out to be sufficient to obtain compositionality (Saraswat *et al.* 1991). Actually it is sufficient to consider a special class of functions: the so-called bounded trace operators. A bounded trace operator (bto) is a closure operator Ψ which is defined on a sublattice of $\mathcal{C}$ of the form

$$\downarrow c = \{x \in \mathcal{C} \mid x \leq c\}$$

where c is an element of $\mathcal{C}$. Moreover, the inverse of Ψ, defined as

$$\Psi^{-1} = (\downarrow c \setminus \Psi) \cup \{c\}$$

is a closure operator on $\downarrow c$ as well. For the definition of the semantic operators on sets of bto's consult (Saraswat et al. 1991).

Saraswat et al. (1991) also obtain full abstraction by applying a sort of saturation condition: if the denotation of a process contains a bto Ψ, then it contains also all the bto's on the same domain which are supersets of Ψ. In the next section we will show the intuition behind this.

5.5.3 Correspondence between reactive sequences and bounded traces operators

The following definition builds a mapping $\mathcal{F}$ from reactive sequences (without termination modes) to bto's.

Definition 5.15 *Let s be a modular sequence. We define the bto $\mathcal{F}(s)$ by induction on the length of s.*

(i) $\mathcal{F}(\lambda)(c) = c$,
(ii) $\mathcal{F}(c^t \cdot s)(c') = c' \wedge \mathcal{F}(s)(c)$,
(iii) $\mathcal{F}(c^a \cdot s)(c') = $ *if* $c \leq c'$ *then* $\mathcal{F}(s)(c)$ *else* c'.

It is easy to prove that this definition gives indeed a bto, with domain $\downarrow con(s)$.

We consider now the opposite function $\mathcal{G}$ which associates a reactive sequence to a bto.

Definition 5.16 *Let Ψ be a bto. We define the reactive sequence $\mathcal{G}(\Psi)$ as follows:*

$$\Psi(true)^t \cdot \Psi^{-1}(\Psi(true))^a \cdot \Psi(\Psi^{-1}(\Psi(true)))^t \ldots$$

The sequence ends when the application of Ψ or Ψ^{-1} doesn't give any new contribution to the store, i.e. when the sequence reaches a fixpoint of Ψ and of Ψ^{-1}.

It is possible to prove that $\mathcal{F}$ is surjective and $\mathcal{G}$ is injective. Furthermore $\mathcal{F}(\mathcal{G}(\Psi)) = \Psi$, whereas $\mathcal{G}(\mathcal{F}(s))$ is in general different from s. We call $\mathcal{G}(\mathcal{F}(s))$ the *canonical representant* of s. The sequences s and $\mathcal{G}(\mathcal{F}(s))$ are equivalent with respect to the saturation conditions: each of them can generate the other by saturation.

Proposition 5.17 *Let s be a reactive sequence. Then*

(i) $s \in Sat(\mathcal{G}(\mathcal{F}(s)))$ *and*
(ii) $\mathcal{G}(\mathcal{F}(s)) \in Sat(s)$.

Another important property is the following: Consider two bto's Ψ and Φ such that $\Psi \subseteq \Phi$. Then $\mathcal{G}(\Phi)$ is a sequence which *asks more* than $\mathcal{G}(\Psi)$

in some points. Note that this corresponds to obtaining $\mathcal{G}(\Phi)$ from $\mathcal{G}(\Psi)$ by application of the saturation condition **C2**. Therefore we have that the application of *Sat* corresponds to the saturation by supersets (upward closure).

Proposition 5.18 *Let Ψ and Φ be two bto's. Then $\Psi \subseteq \Phi$ iff $\mathcal{G}(\Phi) \in Sat(\mathcal{G}(\Psi))$.*

From Propositions 5.17 and 5.18 we derive that the set of saturated sets of reactive sequences and the set of upward closed sets of bto's are isomorphic. This isomorphism is given by the extension of $\mathcal{F}$ on sets. Of course, its inverse is given by the extension of $\mathcal{G}$ on sets. Furthermore, it is possible to show that this isomorphism is preserved by the semantics operators of the language, in the following sense:

Proposition 5.19 *For every operator op of the language, let $\widetilde{op}$ be the corresponding semantics operator of our approach, and $\widehat{op}$ the corresponding semantic operator as defined by Saraswat et al. (1991). Then, for $S_1, \ldots, S_n$ saturated sets of sequences we have*

$$\mathcal{F}(\widetilde{op}(S_1, \ldots, S_n)) = \widehat{op}(\mathcal{F}(S_1), \ldots, \mathcal{F}(S_n)),$$

and for $T_1, \ldots, T_n$ upward closed sets of bto's we have

$$\mathcal{G}(\widetilde{op}(T_1, \ldots, T_n)) = \widehat{op}(\mathcal{G}(T_1), \ldots, \mathcal{G}(T_n)).$$

From Proposition 5.19 we finally derive:

Theorem 5.20 *The fully abstract semantics $\mathcal{A}$ presented in previous sections and the semantics defined by Saraswat et al. (1991) are isomorphic.*

The representation of the semantics in terms of upward closed sets of bto's has the advantage that full abstraction is very easy to prove. Given a bto Ψ belonging to (the denotation of) a process A and not to B, consider a context $C[\ ]$ consisting of the process $\{\Psi^{-1}\}$ in parallel, i.e. $C[\ ] = \{\Psi^{-1}\} \parallel [\ ]$. Then $\Psi \parallel \Psi^{-1} = \Psi \cap \Psi^{-1}$ is a singleton c (the lub of their domain) and it constitutes an observable distinction, in fact $c \in C[A]$ and $c \notin C[B]$. To prove the last fact we use the upward closedness: if by contradiction $c \in C[B]$ then B would contain a bto $\Phi \subseteq \Psi$. But then, since B is upward closed, it would contain Ψ as well.

Acknowledgements

We are grateful to Jan Willem Klop for his helpful comments on previous versions of this paper.

Bibliography

1. Andreoli, J.-M. and Pareschi, R. (1991). Linear Objects: logical processes with built-in inheritance. *New Generation Computing*, 9:445–473.

2. Apt, K. (1990). Introduction to logic programming. In van Leeuwen, J., editor, *Handbook of Theoretical Computer Science*, volume B. Elsevier Science Publishers, Amsterdam.
3. Apt, K., Marchiori, E., and Palamidessi, C. (1992). A theory of first-order built-in's of PROLOG. In Kirchner, H. and Levi, G., editors, *Proc. of the Third Int. Conference on Algebraic and Logic Programming (ALP)*, volume 632 of *Lecture Notes in Computer Science*, pages 69–83. Springer-Verlag, Berlin.
4. Bergstra, J. and Klop, J. (1984). Process algebra for synchronous communication. *Information and Control*, 60(1,3):109–137.
5. Bergstra, J. and Klop, J. (1986). Process algebra: specification and verification in bisimulation semantics. In *Mathematics and Computer Science II*, CWI Monographs, pages 61 – 94. North-Holland, Amsterdam.
6. Bougé, L. (1988). On the existence of symmetric algorithms to find leaders in networks of communicating sequential processes. *Acta Informatica*, 25:179–201.
7. Brookes, S., Hoare, C., and Roscoe, W. (1984). A theory of communicating sequential processes. *Journal of ACM*, 31:499–560.
8. Clark, K. and Gregory, S. (1981). A relational language for parallel programming. Res. Report DOC 81/16, Imperial College, Dept. of Computing, London.
9. Clark, K. and Gregory, S. (1986). PARLOG: parallel programming in logic. *ACM Trans. on Programming Languages and Systems*, 8(1):1–49.
10. de Bakker, J. and Kok, J. (1988). Uniform abstraction, atomicity and contractions in the comparative semantics of Concurrent Prolog. In *Proc. of the International Conference on Fifth Generation Computer Systems*, pages 347–355. Institute for New Generation Computer Technology (ICOT), Tokyo, Japan.
11. de Bakker, J. and Kok, J. (1990). Comparative metric semantics for Concurrent Prolog. *Theoretical Computer Science*, 75(1/2):15–44.
12. de Boer, F. and Palamidessi, C. (1990a). Concurrent logic languages: Asynchronism and language comparison. In Debray, S. and Hermenegildo, M., editors, *Proc. of the North American Conference on Logic Programming*, Series in Logic Programming, pages 175–194. MIT Press, Cambridge, MA.
13. de Boer, F. and Palamidessi, C. (1990b). On the asynchronous nature of communication in concurrent logic languages: A fully abstract model based on sequences. In Baeten, J. and Klop, J., editors, *Proc. of Concur 90*, volume 458 of *Lecture Notes in Computer Science*, pages 99–114. Springer-Verlag, Berlin.
14. de Boer, F. and Palamidessi, C. (1991a). A fully abstract model for concurrent constraint programming. In Abramsky, S. and Maibaum,

T., editors, *Proc. of TAPSOFT/CAAP*, volume 493 of *Lecture Notes in Computer Science*, pages 296–319. Springer-Verlag, Berlin.

15. de Boer, F. and Palamidessi, C. (1991b). Embedding as a tool for language comparison: On the CSP hierarchy. In Baeten, J. and Groote, J., editors, *Proc. of CONCUR 91*, volume 527 of *Lecture Notes in Computer Science*, pages 127 – 141. Springer-Verlag, Berlin.

16. de Boer, F. and Palamidessi, C. (1992). A process algebra for concurrent constraint programming. In Apt, K., editor, *Proc. of the Joint International Conference and Symposium on Logic Programming*, Series in Logic Programming, pages 463–477. MIT Press, Cambridge, MA.

17. de Boer, F. and Palamidessi, C. (1994). Embedding as a tool for language comparison. *Information and Computation* 108(1): 128–157.

18. de Boer, F., Kok, J., Palamidessi, C., and Rutten, J. (1989). Control flow versus logic: a denotational and a declarative model for Guarded Horn Clauses. In Kreczmar, A. and Mirkowska, G., editors, *Proc. of the Symposium on Mathematical Foundations of Computer Science*, volume 379 of *Lecture Notes in Computer Science*, pages 165–176. Springer-Verlag, Berlin.

19. de Boer, F., Kok, J., Palamidessi, C., and Rutten, J. (1991). Semantic models for concurrent logic languages. *Theoretical Computer Science*, 86(1):3 – 33.

20. de Boer, F., Kok, J., Palamidessi, C., and Rutten, J. (1992). From failure to success: Comparing a denotational and a declarative semantics for Horn Clause Logic. *Theoretical Computer Science*, 101(2):239–263.

21. de Boer, F., Kok, J., Palamidessi, C., and Rutten, J. (1993). Non-monotonic concurrent constraint programming. *Proc. of the International Symposium on Logic Programming*, Vancouver, 1993.

22. Dijkstra, E. (1986). *A Discipline of Programming*. Prentice-Hall, Englewood Cliffs, N.J.

23. Ever-Hadani, R. (1987). Operational and denotational linear sequence semantics for full Concurrent PROLOG. Technical report, Department of Computer Science, The Weizmann Institute of Science, Rehovot, Israel. M.Sc. Thesis, 1992.

24. Falaschi, M., Levi, G., and Palamidessi, C. (1984). A synchronization logic: Axiomatics and formal semantics of generalized Horn clauses. *Information and Control*, 60(6):36–69.

25. Falaschi, M., Gabbrielli, M., Marriott, K., and Palamidessi, C. (1993). Compositional analysis for concurrent constraint programming. In *Proc. of the Eight Annual IEEE Symposium on Logic in Computer Science*, pages 210–221. IEEE Computer Society Press, New York.

26. Gabbrielli, M. and Levi, G. (1990). Unfolding and fixpoint semantics for concurrent constraint logic programs. In Kirchner, H. and Wechler,

W., editors, *Proc. of the Second Int. Conf. on Algebraic and Logic Programming*, Lecture Notes in Computer Science, pages 204–216. Springer-Verlag, Berlin.

27. Gaifman, H., Maher, M. J., and Shapiro, E. (1989). Reactive behaviour semantics for concurrent constraint logic programs. In Lusk, E. and Overbeck, R., editors, *North American Conference on Logic Programming*.

28. Gerth, R., Codish, M., Lichtenstein, Y., and Shapiro, E. (1988). Fully abstract denotational semantics for Concurrent Prolog. In *Proc. of the Third Annual IEEE Symposium on Logic In Computer Science*, pages 320–335. IEEE Computer Society Press, New York.

29. Gregory, S. (1987). *Parallel Logic Programming in PARLOG: the Language and its Implementation*. Addison-Wesley, Reading, MA.

30. Harel, D. and Pnueli, A. (1985). On the development of reactive systems. In Apt, K. R., editor, *Logics and Models of Concurrent Systems*. Springer-Verlag, Berlin.

31. Haridi, S. (1990). A logic programming language based on the Andorra model. *New Generation Computing*, 7(2/3):109–125.

32. Haridi, S. and Brand, P. (1988). Andorra PROLOG: an integration of PROLOG and committed choice languages. In *Proc. of the International Conference on Fifth Generation Computer Systems*, pages 745–754. Institute for New Generation Computer Technology (ICOT), Tokyo, Japan.

33. Haridi, S. and Janson, S. (1990). Kernel Andorra PROLOG and its computation model. In Warren, D. and Szeredi, P., editors, *Proc. of the Seventh International Conference on Logic Programming*, pages 31–48.

34. Haridi, S. and Janson, S. (1991). Programming paradigms of the Andorra Kernel Language. In Saraswat V. and Ueda K., editors, *Proc. of the International Logic Programming Symposium*, pages 167–186.

35. Haridi, S., Janson, S., and Palamidessi, C. (1992). Structural tranformational semantics for Andorra Kernel PROLOG. *Future Generation Computer Systems*, 8(4):409–421.

36. Henkin, L., Monk, J., and Tarski, A. (1971). *Cylindric Algebras (Part I)*. North-Holland, Amsterdam.

37. Hoare, C. (1978). Communicating sequential processes. *Communications of the ACM*, 21(8):666–677.

38. Jacquet, J.-M. (1991). *Conclog: a Methodological Approach to Concurrent Logic Programming*, volume 556 of *Lecture Notes in Computer Science*. Springer-Verlag, Berlin.

39. Jacquet, J.-M. and Monteiro, L. (1990). Comparative semantics for a parallel contextual logic programming language. In Debray, S. and

Hermenegildo, M., editors, *Proc. of the North American Conference on Logic Programming*, pages 195–214. MIT Press, Cambridge, MA.

40. Jacquet, J.-M. and Monteiro, L. (1991). Extended Horn clauses: the framework and its semantics. In Baeten, J. and Groote, J., editors, *Proc. of CONCUR 91*, volume 527 of *Lecture Notes in Computer Science*, pages 281–297. Springer-Verlag, Berlin.
41. Jacquet, J.-M. and Monteiro, L. (1992). Communication clauses: towards synchronous communication in contextual logic programming. In Apt, K., editor, *Proc. of the Joint International Conference and Symposium on Logic Programming*, Series in Logic Programming, pages 98–112. MIT Press, Cambridge, MA.
42. Jaffar, J. and Lassez, J.-L. (1987). Constraint logic programming. In *Proc. of ACM Symposium on Principles of Programming Languages*, pages 111–119. ACM, New York.
43. Kliger, S., Yardeni, E., Kahn, K., and Shapiro, E. (1988). The language FCP(:,?). In *Proc. of the International Conference on Fifth Generation Computer Systems*, pages 763–773. Institute for New Generation Computer Technology (ICOT), Tokyo, Japan.
44. Knijnenburg, P. and Kok, J. (1989). A compositional semantics for the finite failures of a language with atomized statements. In *Proc. Computer Science in the Netherlands (CSN 89)*.
45. Lassez, J. L. and Maher, M. J. (1984). Closures and fairness in the semantics of programming logic. *Theoretical Computer Science*, 29:167–184.
46. Levi, G. and Palamidessi, C. (1985). The declarative semantics of logical read-only variables. In *Proc. of the Third IEEE Symposium on Logic Programming*, pages 128–137. IEEE Computer Society Press, New York.
47. Levi, G. and Palamidessi, C. (1987). An approach to the declarative semantics of synchronization in logic languages. In Lassez, J.-L., editor, *Proc. of the Fourth International Conference on Logic Programming*, pages 877–893. MIT Press, Cambridge, MA.
48. Levi, G., Martelli, M., and Palamidessi, C. (1990). Failure and success made symmetric. In Debray, S. and Hermenegildo, M., editors, *Proc. of the North American Conference on Logic Programming*, Series in Logic Programming, pages 3–22. MIT Press, Cambridge, MA.
49. Levy, J. (1988). *Concurrent PROLOG and Related Languages*. PhD thesis, Department of Computer Science, The Weizmann Institute of Science, Rehovot, Israel.
50. Lloyd, J. (1987). *Foundations of Logic Programming* (2nd edn). Springer-Verlag, Berlin.
51. Maher, M. J. (1987). Logic semantics for a class of committed-choice

programs. In Lassez, J.-L., editor, *Proc. of the Fourth International Conference on Logic Programming*, Series in Logic Programming, pages 858–876. MIT Press, Cambridge, MA.

52. Milner, R. (1980). *A Calculus of Communicating Systems*, volume 92 of *Lecture Notes in Computer Science*. Springer-Verlag, New York.

53. Montanari, U. and Rossi, F. (1991). True concurrency semantics for concurrent constraint programming. In Saraswat, V. and Ueda, K., editors, *Proc. of the International Logic Programming Symposium*, Series in Logic Programming, pages 694–716. MIT Press, Cambridge, MA.

54. Monteiro, L. (1981). An extension to Horn clause logic allowing the definition of concurrent processes. In *Proc. of Formalization of Programming Concepts*, volume 107 of *Lecture Notes in Computer Science*, pages 401–407. Springer-Verlag, Berlin.

55. Monteiro, L. (1982). A Horn clause-like logic for specifying concurrency. In *Proc. of the First Int. Conf. on Logic Programming*, pages 1–8.

56. Palamidessi, C. (1990). Algebraic properties of idempotent substitutions. In Paterson, M., editor, *Proc. of the 17th International Colloquium on Automata, Languages and Programming (ICALP)*, volume 443 of *Lecture Notes in Computer Science*, pages 386–399. Springer-Verlag, Berlin.

57. Panangaden, P., Saraswat, V., Scott, P., and Seely, R. (1992). A hyperdoctrinal view of constraint systems. In *Proceedings of the REX Workshop "Semantics: Foundations and Applications"*, Lecture Notes in Computer Science. Springer-Verlag, Berlin.

58. Pereira, L. and Nasr, R. (1984). Delta-PROLOG: A distributed logic programming language. In *Proc. of the International Conference on Fifth Generation Computer Systems*, pages 283–291. Institute for New Generation Computer Technology (ICOT), Tokyo, Japan.

59. Plotkin, G. (1981). A structured approach to operational semantics. Technical Report DAIMI FN-19, Computer Science Department, Aarhus University.

60. Safra, S. (1986). Partial evaluation of concurrent PROLOG and its implications. Technical Report CS86-24, Dept. of Computer Science, The Weizmann Institute of Science, Rehovot, Israel. M.Sc. Thesis.

61. Saraswat, V. (1986). Problems with concurrent PROLOG. Technical Report CMU-CS-86-100, Computer Science Department, Carnegie-Mellon University.

62. Saraswat, V. (1987a). The concurrent logic programming language CP: definition and operational semantics. In *Conference Record of the Fourteenth Annual ACM Symposium on Principles of Programming Languages*, pages 49–63. ACM, New York.

63. Saraswat, V. (1987b). GHC: operational semantics, problems and relationship with CP($\downarrow, |$). In *Proc. of the Fourth IEEE Symposium on Logic*

Programming, pages 347–358. IEEE Computer Society Press, New York.
64. Saraswat, V. (1989). *Concurrent Constraint Programming*. PhD thesis, Carnegie-Mellon University. In ACM distinguished dissertation series. MIT Press, Cambridge, MA.
65. Saraswat, V. (1992). The category of constraint systems is cartesian-closed. In *Proc. of the Seventh Annual IEEE Symposium on Logic in Computer Science*. IEEE Computer Society Press, New York.
66. Saraswat, V. and Lincoln, P. (1992). Higher-order, linear concurrent constraint programming. Technical report, Xerox PARC.
67. Saraswat, V. and Rinard, M. (1990). Concurrent constraint programming. In *Proc. of the seventeenth ACM Symposium on Principles of Programming Languages*, pages 232–245. ACM, New York.
68. Saraswat, V., Rinard, M., and Panangaden, P. (1991). Semantics foundations of concurrent constraint programming. In *Proc. of the eighteenth ACM Symposium on Principles of Programming Languages*. ACM, New York.
69. Scott, D. (1982). Domains for denotational semantics. In *Proc. of ICALP*.
70. Shapiro, E. (1983). A subset of Concurrent PROLOG and its interpreter. Technical Report TR-003, Institute for New Generation Computer Technology (ICOT), Tokyo.
71. Shapiro, E. (1986). Concurrent PROLOG: A progress report. *Computer*, 19(8):44–58.
72. Shapiro, E. (1988). *Concurrent PROLOG: Collected Papers*. Number Vol. 1-2 in Series in Logic Programming. MIT Press, Cambridge, MA.
73. Shapiro, E. (1989). The family of concurrent logic programming languages. *ACM Computing Surveys*, 21(3):412–510.
74. Shapiro, E. (1991). Separating concurrent languages with categories of language embeddings. In *Proceedings of the 23rd Annual ACM Symposium on Theory of Computing*, pages 198–208.
75. Shapiro, E. (1992). Embeddings among concurrent programming languages. In Cleaveland, W., editor, *Proc. of CONCUR 92*, volume 630 of *Lecture Notes in Computer Science*, pages 486–503. Springer-Verlag.
76. Takeuchi, A. and Furukawa, K. (1986). Parallel logic programming languages. In Shapiro, E., editor, *Proc. of the Third International Conference on Logic Programming*, volume 225 of *Lecture Notes in Computer Science*, pages 335–349. Springer-Verlag, Berlin.
77. Ueda, K. (1987). Guarded Horn Clauses. In Shapiro, E. Y., editor, *Concurrent PROLOG: Collected Papers*, Series in Logic Programming. MIT Press, Cambridge, MA.
78. Ueda, K. (1988). Guarded Horn Clauses, a parallel logic programming

language with the concept of a guard. In Nivat, M. and Fuchi, K., editors, *Programming of Future Generation Computers*, pages 441–456. North Holland, Amsterdam.

79. van Emden, M. and de Lucena, G. (1982). Predicate logic as language for parallel programming. In Clark, K. and Tarnlund, S., editors, *Logic Programming*, pages 189–198. Academic Press, London.

80. Yang, R. (1986). *A Parallel Logic Programming Language and its implementation*. PhD thesis, Keio University.

81. Yang, R. and Aiso, H. (1986). P-PROLOG: a parallel language based on exclusive relation. In Shapiro, E., editor, *Proc. of the Third International Conference on Logic Programming*, volume 225 of *Lecture Notes in Computer Science*, pages 255–269. Springer-Verlag, Berlin.

3
Formal Bases for Dataflow Analysis of Logic Programs

Saumya K. Debray
The University of Arizona, Tucson

Abstract

Compile-time dataflow analysis finds a variety of applications, such as compile-time optimization, program parallelization, testing, program transformation and specialization, and program verification. This chapter gives a broad overview of the field of dataflow analysis of logic programs. Starting with a general discussion of abstract interpretation, a formal framework that underlies much of the work on static program analysis, it discusses some of the wide variety of frameworks that have been proposed for the analysis of logic programs, considers the relationship between complexity and precision of analysis, and examines some specific analyses that have been proposed in the literature.

1 Introduction

As software technology matures, programmers tend to rely more and more on programs that manipulate other programs, such as compilers, interpreters, program verifiers, type checkers, program transformation systems, program optimizers and parallelizers, and so on. An important component of such programs is the ability to analyze and reason about the behavior of the programs they manipulate. Now program manipulation tools are usually expected to terminate for all input programs, regardless of whether these programs would terminate when executed. Unfortunately, it turns out that most "interesting" program properties are undecidable (e.g., consider Rice's theorem), which implies that for most such program properties, there is no algorithm that will analyze programs exactly—that is, say that a program has a certain property if and only if that program does, indeed, have that property—and that is also guaranteed to terminate on all inputs. This necessarily implies that "static" analyses, i.e., analyses that are carried out without actually executing the program being analyzed, must involve some kind of finitely computable approximation to the properties of interest.

Many interesting questions arise once we accept the fact that we must be content, at compile time, with approximate descriptions of a program's runtime behavior. These include, for example, questions about different kinds of approximations (e.g., those that predict what "may" happen when a program is executed, compared to those that predict what "must" happen); frameworks for such analyses; their relationship to the actual semantics of the language under consideration; algorithms for dataflow analysis; relationships between the cost and precision of analysis; and applications of dataflow analysis. In this paper, we describe some of the work that attempts to address these questions and give a broad overview of the field of dataflow analysis of logic programs. This is a very active field, with a large and rapidly growing body of literature, and a detailed and exhaustive account would be well beyond the scope of this paper. We have tried, instead, to cover a sampling of topics in order to give the reader a sense of some of the issues in dataflow analysis of logic programs. Such a sampling must, by its very nature, be subjective and imperfect, and the inclusion or omission of any particular body of work in the discussion that follows should not be construed as any form of judgement on the quality of that work.

The remainder of this paper is organized as follows: Section 2 defines basic notions and terminology; Section 3 discusses the theory of abstract interpretation; Section 4 discusses frameworks for the analysis of logic programs; Section 5 addresses complexity and efficiency issues; Section 6 discusses a variety of specific analyses that have been proposed for logic programs; Section 7 briefly describes some proposals for the analysis of concurrent programs; Section 9 discusses the needs of "users" of dataflow information; and Section 10 considers an approach to the analysis of modular programs.

2 Definitions

A binary relation R on a set S is said to be a *partial order* if R is reflexive (i.e., xRx for all $x \in S$), antisymmetric (i.e., for all $x, y \in S$, xRy and yRx implies $x = y$), and transitive (i.e., for all $x, y, z \in S$, if xRy and yRz then xRz). A relation R over a set S is said to be a *total order* if R is a partial order, and additionally, if every pair of elements in S is related under R, i.e., if xRy or yRx for every pair of elements $x, y \in S$. A set S over which there is a partial order R is said to be a *partially ordered set*, or *poset*: a poset S with partial order $\leq$ is sometimes written $\langle S, \leq \rangle$. Given a poset $\langle S, \leq \rangle$ and $X \subseteq S$, an element $s \in S$ is said to be an *upper bound* on X if $x \leq s$ for every $x \in X$; dually, s is a *lower bound* on X if $s \leq x$ for every $x \in X$. An upper bound s on a set $X \subseteq S$ is said to be the *least upper bound* if $s \leq s'$ for any $s' \in S$ that is an upper bound on X; dually, s is the *greatest lower bound* of X if $s' \leq s$ for any $s' \in S$ that is a lower bound on X. Given a poset $\langle S, \leq \rangle$ and $X \subseteq S$, there may or may not be a least upper bound or a greatest lower bound on X; however, it follows from the fact that $\leq$ is a partial order, and therefore antisymmetric, that the least upper bound of X, if it exists, is unique (and similarly with the greatest lower bound). A *lattice* is a poset $\langle S, \leq \rangle$ where every pair of elements (and, by extension, every finite subset of S) has a least upper bound and a greatest lower bound; a lattice is said to be *complete* if for any $X \subseteq S$, the least upper bound and the greatest lower bound of X exists. Note that this implies that every complete lattice has a least element and a greatest element.

Given a poset $\langle S, \leq \rangle$, a set $X \subseteq S$ is said to be a *chain* if X is totally ordered by $\leq$. Given posets $\langle R, \sqsubseteq \rangle$ and $\langle S, \leq \rangle$, a function $f : R \longrightarrow S$ is said to be *monotone* if for every $x, y \in R$, if $x \sqsubseteq y$ then $f(x) \leq f(y)$; f is said to be *continuous* if for any chain $X \subseteq R$, $f(\sqcup X) = \vee \{f(x) \mid x \in X\}$, where $\sqcup X$ denotes the least upper bound of X, and $\vee Y$ denotes the least upper bound of any $Y \subseteq S$. Given a set S and a function $f : S \longrightarrow S$, an element $s \in S$ is said to be a *fixpoint* of f if $f(s) = s$. The following result is due to Tarski (1955):

Theorem 2.1 *If L is a complete lattice and $f : L \longrightarrow L$ is monotone, the fixpoints of f form a complete lattice.*

In particular, this implies that a monotone function f on a complete lattice L has a *least fixpoint* $lfp(f)$. Furthermore, if f is continuous, then $lfp(f) = \sqcup \{f^i(\bot)\}$, where $\sqcup$ denotes the least upper bound operator, $\bot$ is the least element of L, and for any $x \in L$, $f^k(x)$ denotes the result of k applications of f to x.

A binary relation R over a set S is said to be an *equivalence relation* if R is symmetric (i.e., for all $x, y \in S$, xRy implies yRx), reflexive, and transitive. Given an equivalence relation R on a set S, and any element x of S, the set $\{y \in S \mid yRx\}$ is said to be the *equivalence class* of x; the set of equivalence classes **S** obtained from an equivalence relation over a set S partitions S, i.e., $\cup \mathbf{S} = S$ and for any pair of elements $X, Y \in \mathbf{S}$, if $X \neq Y$ then $X \cap Y = \emptyset$. Let $\mathcal{A}$ denote a set A together with a collection of operations F over A, and let $\equiv$ be an equivalence relation over A, then the *quotient* of $\mathcal{A}$ with respect to $\equiv$, written $\mathcal{A}/\equiv$, consists of the set of equivalence classes of A induced by $\equiv$, together with a collection of operations F' corresponding to F such that for every operation f in F there is an operation $f' \in F'$ defined as:

$$f'([x_1], \ldots, [x_n]) = [f(x_1, \ldots, x_n)]$$

where $[x]$ denotes the equivalence class of an element x of A. In contexts where the operations of interest are clear, the quotient is sometimes written $A/\equiv$.

Most logic programming languages are based on a subset of the first order predicate calculus known as Horn clause logic. Such a language has a countably infinite set of variables, and countable sets of function and predicate symbols, these sets being mutually disjoint. Without loss of generality, we assume that with each function symbol f and each predicate symbol p is associated a unique natural number n, referred to as the *arity* of the symbol; f and p are said to be n-ary symbols, and written f/n and p/n respectively. A 0-ary function symbol is referred to as a constant.

A term in such a language is either a variable, or a constant, or a compound term $f(t_1, \ldots, t_n)$ where f is an n-ary function symbol and the t_i are terms. A literal is either an atom $p(t_1, \ldots, t_n)$, where p is an n-ary predicate symbol and $t_1, \ldots, t_n$ are terms, or the negation of an atom; in the first case the literal is said to be positive, in the second case it is negative. A clause is the disjunction of a finite number of literals, and is said to be Horn if it has at most one positive literal. A Horn clause with exactly one positive literal is referred to as a *definite clause*. The positive literal in a definite clause is its *head*, and the remaining literals constitute its *body*. A predicate definition consists of a finite number of definite clauses, all of whose heads have the same predicate symbol; a goal is a set of negative literals. A logic program consists of a finite set of predicate definitions. The set of variables occurring in a term (literal, goal) t is denoted by $vars(t)$. In this paper, we adhere to the syntax of Edinburgh PROLOG and write a definite clause as

$$p :- q_1, \ldots, q_n.$$

read declaratively as "p if q_1 and ... and q_n." Names of variables begin with upper case letters, while names of non-variable (i.e. function and predicate) symbols begin with lower case letters.

A substitution is an idempotent mapping from a finite set of variables to terms. A substitution σ_1 is said to be *more general* than a substitution σ_2 if there is a substitution θ such that $\sigma_2 = \theta \circ \sigma_1$. Two terms t_1 and t_2 are said to be *unifiable* if there exists a substitution σ such that $\sigma(t_1) = \sigma(t_2)$; in this case, σ is said to be a *unifier* for the terms. If two terms t_1 and t_2 have a unifier, then they have a *most general unifier* that is unique up to variable renaming.

The operational behavior of logic programs can be described by means of *SLD-derivations*. An SLD-derivation for a goal G with respect to a program P is a sequence of goals $G_0, \ldots, G_i, G_{i+1}, \ldots$ such that $G_0 = G$, and if $G_i =$ "$a_1, \ldots, a_n$", then $G_{i+1} = \theta(a_1, \ldots, a_{i-1}, b_1, \ldots, b_m, a_{i+1}, \ldots, a_n)$ such that $1 \leq i \leq n$; $b :- b_1, \ldots, b_m$ is an alphabetic variant of a clause in P and has no variable in common with any of the goals $G_0, \ldots, G_i$; and θ is the most general unifier of a_i and b. The goal G_{i+1} is said to be obtained from G_i by means of a *resolution* step, and a_i is said to be the *resolved atom*. Intuitively, each resolution step corresponds to a procedure call. Let $G_0, \ldots, G_n$ be an SLD-derivation for a goal G in a program P, and let θ_i be the unifier obtained when resolving the goal G_i to obtain G_{i+1}, $0 \leq i < n$; if this derivation is finite and maximal, i.e. one in which it is not possible to resolve the goal G_n with any of the clauses in P, then this corresponds to a terminating computation for G: in this case, if G_n is the empty goal then the computation is said to *succeed* with answer substitution θ, where θ is the substitution obtained by restricting the substitution $\theta_n \circ \cdots \circ \theta_0$ to the variables occurring in G. If G_n is not the empty goal, then the computation is said to *fail*. If the derivation is infinite, the computation does not terminate.

In practice, logic programming languages usually allow programs where clause bodies may contain negated literals as well as positive literals. Such programs are called *normal* programs. Operationally, negated literals are evaluated using the "negation as failure" rule: a negated literal $\neg L$ is selected for execution only if it is ground; the execution succeeds if every proof tree for L fails in a finite number of resolution steps, fails if any proof tree for L succeeds in a finite number of steps, and is nonterminating otherwise. The extension of SLD-resolution to deal with negation in this way is referred to as SLDNF-resolution.

Let $p(\bar{t})$ be the resolved atom in some SLD-derivation of a goal G in a program P, then we say that $p(\bar{t})$ is a *call* that arises in the computation of G in the program. If the goal $p(\bar{t})$ can succeed with answer substitution θ, then we also say that it can succeed with its arguments bound to $\theta(\bar{t})$.

3 Abstract Interpretation

3.1 Overview

Much of the work on dataflow analysis of logic programs is formulated in terms of a general framework for program analysis called *abstract interpretation* (Cousot 1981, Cousot and Cousot 1977, 1979, 1992). In very general terms, the basic idea here is to set up and solve a system of equations called *dataflow equations*. A dataflow equation describes properties that hold at one point in a program in terms of properties at other program points. Differences between various analyses arise from the particular properties under consideration, how these properties are represented and manipulated, the way in which equations are solved,[12] even the notion of "program point."

For now, it suffices to have an intuitive notion of a "program point" as being a point, roughly akin to a program counter value, where we can take a "snapshot" of the state of a computation. Given a set of program points for a program, we can then say that a point p_1 is *before* another point p_2 (equivalently, p_2 is *after* p_1) if execution always reaches p_1 before it reaches p_2. Now consider a dataflow analysis where the dataflow equations are of the form

---

[12] Some authors take the position that the framework of abstract interpretation necessarily relies on the use of iterative techniques for solving recursive dataflow equations. It is not clear that the foundational papers on abstract interpretation mandate this, however, so we will not presuppose the use of iterative techniques.

$$\pi_0(p_0) = f(\pi_1(p_1), \ldots, \pi_n(p_n))$$

for some function f, where $p_0, p_1, \ldots, p_n$ are program points and $\pi_0, \pi_1, \ldots, \pi_n$ are properties that hold at program points. If the point p_0 is after each of the points $p_1, \ldots, p_n$ then dataflow information is being propagated "forward" along the direction of control flow, and the analysis is called a *forward analysis*. On the other hand, if p_0 is before each of the points $p_1, \ldots, p_n$ then dataflow information is being propagated "backward" against the direction of control flow, and the analysis is called a *backward analysis*. An example of a forward analysis in the context of traditional compiler analyses for imperative languages is "reaching definitions analysis", while "live variable analysis" is an example of a backward analysis.

By analogy with the notions of forward and backward analyses, there are two kinds of dataflow analyses of logic programs, called *top-down* and *bottom-up* analyses. The fundamental difference between these arises from the underlying "concrete semantics" used. Top-down analyses are based on semantics that describe "top-down" execution strategies: such semantics include operational descriptions, such as SLD-resolution (van Emden and Kowalski 1976), and various denotational descriptions of the semantics of PROLOG (see, for example, Debray and Mishra 1988, Jones and Mycroft 1984, Jones and Søndergaard 1987). By contrast, bottom-up analyses are based on model-theoretic semantics of logic programs (e.g., see Falaschi *et al.* (1989), van Emden and Kowalski 1976). While the distinction between the underlying semantics motivated the development of these two kinds of analyses, however, the distinction is not as fundamental as it seems, since by appropriate choice of the notion of "program point," top-down and bottom-up analyses can be seen as nothing but forward and backward analyses in the usual sense. To see this, consider the analysis of a clause

$$H :- B_1, \ldots, B_n.$$

A top-down analysis mimics a top-down execution strategy such as SLD-resolution: first, it considers unification of the goal with the head H, after which the body literals $B_1, \ldots, B_n$ are processed according to the execution strategy. Most top-down analyses attach two "program points" to each literal: one corresponding to the "entry" to that literal, i.e., the point where that literal is about to be called (i.e., resolved with the head of a clause), and another corresponding to the "return" from that literal, i.e., where that literal and all the literals in the SLD-tree below it have been resolved away. Since the body literals of a clause are considered in the order that they would be executed at runtime, information flow is in the direction of control flow, so top-down analyses are nothing but forward analyses. Now suppose we ignore the entry points for literals, and focus only on the return points: then, since information about the return point for a literal $p(\bar{t})$ must be obtained from information about return points of literals in the bodies of clauses defining p, information flow is opposite to the direction of control flow, so bottom-up analyses can be seen as a restricted kind of backward analyses (restricted in that information at "entry points" is not considered).

Dataflow analyses make predictions about the runtime behavior of a program. Obviously, any such prediction must apply to *all possible executions* of the program, for otherwise it would not be very useful. We can imagine dealing with this by "collecting together" information about all possible executions of the program (on the inputs of interest), then extracting from this the information we want ("does property π hold at program point p in every possible execution of the program?"). Of course, we would probably not want to actually try to collect information about all possible executions of a program, since this could be very time-consuming, and potentially non-terminating—however, we can use the formal semantics of the program to specify, mathematically, the information we would get at each program point if this were done. The resulting mathematical objects simply specify, at each program point, the set of "states" or "snapshots" of the program when execution reaches that point, collected over all possible executions. Moreover, the primitive operations of the language under consideration can be considered to operate over these sets rather than over their original domains—for example, while the "usual semantics" of an addition operation "+" might be a function that maps pairs of numbers to numbers, when we consider all possible executions the domain becomes sets of pairs of numbers rather than individual numbers, and these are mapped to sets of numbers. The result of dealing with "all possible executions" in this way yields a domain, and an interpretation of the operations of the language over this domain, that can itself be thought of as a semantics for the program. It is referred to as the *collecting semantics* (Cousot and Cousot (1977) call it the "static semantics"). Since we are discussing compile-time analyses, we will be concerned primarily with collecting semantics in the remainder of this chapter. The domain of the collecting semantics is usually referred to as the *concrete domain* (to distinguish it from the "abstract domain," which is obtained by abstracting away details about the execution of the program that are not relevant to the particular analysis under consideration), and written $\mathcal{D}_{conc}$.

Intuitively, each element of the concrete domain represents a set of possibilities that may be realized during a computation. If U represents the universe of all possible outcomes, therefore, we can think of $\mathcal{D}_{conc}$ as the partially ordered set $\langle \mathcal{D}_{conc}, \subseteq \rangle$, which forms a complete lattice. In general, of course, semantic definitions may include domains other than powersets, so the concrete domain is generally considered to be an arbitrary complete lattice $\langle \mathcal{D}_{conc}, \sqsubseteq \rangle$. The greatest element $\top$ of $\mathcal{D}_{conc}$ denotes "all possibilities", which conveys no information, while the least element $\bot$ denotes "no possibilities", which represents an impossibility (i.e., an error or failure of execution). Note that this is the opposite of the usual ordering used for domains in denotational semantics, where $\bot$ denotes no information and $\top$ represents an error.

One can distinguish between two general approaches to program analysis. In the *Galois connection* approach, the dataflow equations are set up over a domain of *descriptions*, each element of which "describes" a set of possible "concrete values", i.e., values that could arise during an actual computation. In the *widening/narrowing* approach, on the other hand, the dataflow equations are set up directly over concrete values. The two can also be combined to yield a hybrid approach where widening/narrowing is used in the context of a Galois connection approach using a domain of descriptions. Finally, a set of dataflow equations need not have a unique solution, and it is possible to carry out dataflow analyses by computing the largest solution for a set of dataflow equations rather than the smallest solution.

3.2 The Galois connection approach

The general idea in the Galois connection approach to abstract interpretation is to define a domain of descriptions and symbolically execute the program over this domain. The domain of descriptions is usually referred to as the *abstract domain*. Each element of the abstract domain "describes" a set of values in the concrete domain. To simulate the computation over the abstract domain, it is necessary to specify how each primitive operation involved in the computation is to be dealt with in the abstract domain. Thus, corresponding to each "concrete operation" f of the language under consideration, we must specify an "abstract operation" $\widehat{f}$ over the abstract domain.

The abstract domain: Elements of the abstract domain $\mathcal{D}_{abs}$ "describe" (in a sense that will be made precise) sets of elements in the concrete domain $\mathcal{D}_{conc}$. What should $\mathcal{D}_{abs}$ look like? This question can be answered by considering what properties $\mathcal{D}_{abs}$ should have in order to allow us to reason about various properties of the analyses of interest in a convenient way:

(1) To facilitate reasoning about fixpoints, it is convenient if $\mathcal{D}_{abs}$ is a complete lattice.

(2) The *meaning* of each element of $\mathcal{D}_{abs}$, i.e., the set of "computational possibilities" it describes, should be well defined. Thus, there should be a function

$$conc : \mathcal{D}_{abs} \longrightarrow \mathcal{D}_{conc}.$$

This function is generally referred to as the "concretization" function. It makes little sense to have redundant descriptions in $\mathcal{D}_{abs}$, i.e., many different descriptions for the same set of concrete computational possibilities. Thus, *conc* should be one-one.

(3) The abstract domain $\mathcal{D}_{abs}$ should be "structurally similar" to $\mathcal{D}_{conc}$, in the sense of the ordering "provides more information" on their elements. Suppose the abstract and concrete domains are partially ordered sets $\langle \mathcal{D}_{abs}, \leq \rangle$ and $\langle \mathcal{D}_{conc}, \sqsubseteq \rangle$ respectively. Given two elements d_1, d_2 in $\mathcal{D}_{abs}$, suppose

$$conc(d_1) \sqsubseteq conc(d_2),$$

i.e., the image of d_1 under *conc*, which represents the set of computational possibilities that d_1 denotes, provides more information than the image of d_2. It is reasonable to insist that this should be the case if and only if d_1 provides more information than d_2, i.e., $d_1 \leq d_2$. In other words, *conc* should be monotone.

In addition, it may not be unreasonable to require that each element in the concrete domain should have a unique "best" (i.e., most precise) description in the abstract domain. Strictly speaking, this requirement is not necessary if, for example, we are interested only in soundness, and some authors have described dataflow analyses of logic programs without presupposing the existence of such "best" descriptions—see, for example, Debray (1988) and Marriott and Søndergaard (1992). However, if there is no unique "best" description for some concrete domain element d, then we have to determine which abstract domain element should be used to describe d. Cousot and Cousot (1979) show that in this case, the description that should be chosen (if we want the final result of the analysis to be as good as possible) depends on the program being analyzed, and in general cannot be predicted beforehand; however, each element of $\mathcal{D}_{conc}$ has a unique best description in $\mathcal{D}_{abs}$ if and only if $\mathcal{D}_{abs}$ is a *Moore family*, where a Moore family is defined as follows (Birkhoff 1940):

Definition 3.1 *Let $\langle L, \sqsubseteq \rangle$ be a complete lattice with greatest element $\top$, then $S \subseteq L$ is a Moore family if and only if (i) $\top \in S$; and (ii) for every $T \subseteq S$, $\sqcap T \in S$.*

In this case, we can define a function $abs : \mathcal{D}_{conc} \longrightarrow \mathcal{D}_{abs}$, usually referred to as the "abstraction function", that maps each element of the concrete domain to the abstract domain element that best describes it. Arguing as above, it is not difficult to see that like *conc*, the function *abs* must also be monotone.

If the abstract and concrete domains are such that both abstraction and concretization functions are defined, then we can say more about the relationships between these domains. First, it seems reasonable to require that each element of $\mathcal{D}_{abs}$ be a description of at least one element of $\mathcal{D}_{conc}$ (otherwise, we could delete such points in $\mathcal{D}_{abs}$, which serve no useful purpose, with no difference to our analyses): this requires that *abs* be an onto function. Finally, since there may be some loss of information in using an element of $\mathcal{D}_{abs}$ to describe an element of $\mathcal{D}_{conc}$ (in the sense that, given some element $d \in \mathcal{D}_{conc}$, the "best" description $abs(d)$ may not describe exactly the set d, but rather an approximation to it), it is reasonable to require that

(1) the description be sound, i.e., $conc(abs(x)) \sqsupseteq x$ for all $x \in \mathcal{D}_{conc}$; and

(2) all the loss of information should happen at once, i.e., $abs(conc(x)) = x$ for all $x \in \mathcal{D}_{abs}$.

If the functions abs and $conc$ satisfy these two requirements, they are said to be *adjoint*, and define a *Galois connection* between $\mathcal{D}_{conc}$ and $\mathcal{D}_{abs}$.[13] (Our discussion here has followed the original formulation of abstract interpretation due to Cousot and Cousot (1977), where both abstraction and concretization functions are required to exist. However, various researchers have considered other "frameworks" for abstract interpretation where this requirement is relaxed in various ways. A discussion of such frameworks, their advantages and disadvantages, and a description of a general framework that encompasses these various approaches may be found in Marriott (1993).)

The analysis must also be able to account for the effects of concrete operations in the program: for this, it is necessary to "simulate" the effects of operations in the language under consideration—in general, it suffices to deal with the primitive operations of the language. Corresponding to each primitive operation $f : \mathcal{D}_{conc} \longrightarrow \mathcal{D}_{conc}$, therefore, there must be an "abstract operation" $\widehat{f} : \mathcal{D}_{abs} \longrightarrow \mathcal{D}_{abs}$ that simulates the effects of f over the abstract domain. Each such abstract function must be at least *locally correct*, i.e., correctly describe the effects of a single f operation. In other words, suppose $x \in \mathcal{D}_{conc}$ represents some set of concrete domain values that might actually be encountered at runtime: if the operation f is executed, the resulting set of values that could be realized at runtime is $f(x)$. Now at analysis time, all we have is a description $abs(x)$, and if we execute the abstract operation $\widehat{f}$ on this, the resulting description $\widehat{f}(abs(x))$ must correctly describe the actual runtime possibilities, given by $f(x)$. Of course, it may happen that there is some loss of precision in general, so the description $\widehat{f}(abs(x))$ may not be as good a description of $f(x)$ as possible, but it must be sound. This requirement can be expressed as

$$f(x) \sqsubseteq conc(\widehat{f}(abs(x))) \quad \text{for all } x \in \mathcal{D}_{conc}. \tag{$*$}$$

---

[13]Strictly speaking, for a Galois connection between $\mathcal{D}_{conc}$ and $\mathcal{D}_{abs}$ it suffices to have $\forall x \in \mathcal{D}_{abs} \, \forall y \in \mathcal{D}_{conc} . abs(y) \leq x \Leftrightarrow y \sqsubseteq conc(x)$, which is equivalent to requiring condition (1) together with the condition $x \leq abs(conc(x))$ for all $x \in \mathcal{D}_{abs}$. If the stronger condition (2) is satisfied, the resulting structure is sometimes referred to as a *Galois insertion*.

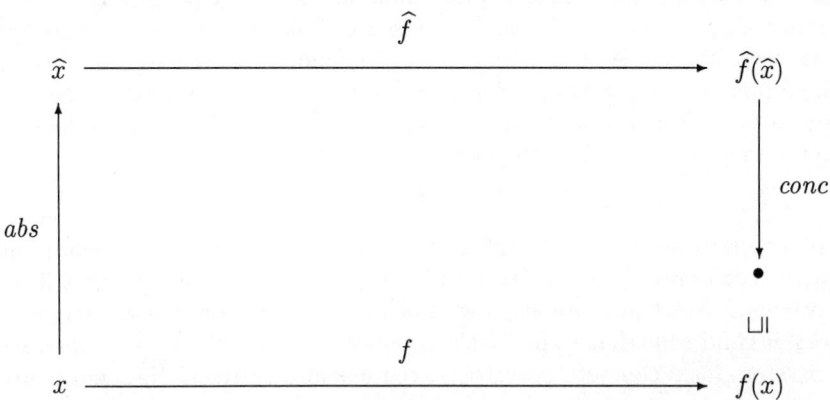

FIG. 1. Correctness of abstract operations.

This is represented pictorially in Figure 1.

This local correctness condition implies that a single execution step over the concrete domain is correctly described by a single computation step over the abstract domain. From this, it is not difficult to show that if the local correctness condition is satisfied for each primitive operation of the language, then for any finite n, the result of n execution steps over the concrete domain is correctly described by the result of n computation steps over the abstract domain. Typically, however, we are interested in a correct description of the least fixpoint of some operation(s) over the concrete domain, and it may happen that this least fixpoint is not attainable in finitely many steps. It might appear, then, that some form of fixpoint induction might be necessary to reason about the "global correctness" of program analyses, with various complications such as the need for additional assumptions about admissibility of predicates. As the following theorem shows, such complications do not arise in the Galois connection approach:

Theorem 3.2. (Cousot & Cousot 1977) *Consider an abstract interpretation with concrete domain $\langle \mathcal{D}_{conc}, \sqsubseteq \rangle$ and abstract domain $\langle \mathcal{D}_{abs}, \leq \rangle$ that are complete lattices, with abs $: \mathcal{D}_{conc} \longrightarrow \mathcal{D}_{abs}$ and conc $: \mathcal{D}_{abs} \longrightarrow \mathcal{D}_{abs}$ monotone and adjoint. Then, if the abstract operator $\widehat{f} : \mathcal{D}_{abs} \longrightarrow \mathcal{D}_{abs}$ is locally consistent with the concrete operator $f : \mathcal{D}_{conc} \longrightarrow \mathcal{D}_{conc}$, i.e., $f(x) \sqsubseteq conc(\widehat{f}(abs(x)))$ for every $x \in \mathcal{D}_{conc}$, then $lfp(f) \sqsubseteq conc(lfp(\widehat{f}))$ and $gfp(f) \sqsubseteq conc(gfp(\widehat{f}))$.*

Finally, there is the question of the design of abstract interpretations. Typically, the concrete domain $\mathcal{D}_{conc}$ is fixed by the programming language under consideration, and we select an abstract domain $\mathcal{D}_{abs}$ (and therefore the abstraction and concretization functions $abs : \mathcal{D}_{conc} \longrightarrow \mathcal{D}_{abs}$ and $conc : \mathcal{D}_{abs} \longrightarrow \mathcal{D}_{conc}$) based on the particular program properties we are interested in. At this point, all that remains is to define the abstract operations corresponding to the concrete primitive operations of the language in such a way as to ensure local correctness. This can be done in many ways, e.g., an abstract operation $\widehat{f}$ defined as

$$\widehat{f}(x) = \top \quad \text{for all } x \in \mathcal{D}_{conc} \quad [14]$$

i.e., one that always "gives up" and returns no information, is sound, but rather conservative and generally not very useful. It turns out that if we have an abstract interpretation with abstraction and concretization operators abs and $conc$ that form a Galois connection, then of all sound abstract operators for a concrete operator f, there is an "optimal" (i.e., most precise) operator $\widehat{f}$ given by

$$\widehat{f} = abs \circ f \circ conc.$$

(This is, of course, only a specification of $\widehat{f}$—we would probably choose to implement it in a more efficient way.) Unfortunately, optimality of abstract operators is not preserved by composition. In other words, let $\widehat{f}$ and $\widehat{g}$ be the optimal abstract operators corresponding to concrete operators f and g, then in general $\widehat{f} \circ \widehat{g}$ is not necessarily optimal for $f \circ g$. For a particular application, of course, the choice of how to define the abstract operations might be governed by considerations of efficiency and precision—as we will see later, there is a strong correlation between the computational cost of an analysis and its precision.

In summary, a *concrete interpretation*, specifying the actual runtime behavior of a program, can be specified in terms of a concrete domain, which is a complete lattice $\langle \mathcal{D}_{conc}, \sqsubseteq \rangle$, and a set $\mathcal{F}$ of "concrete operations" of the language, which are functions over $\mathcal{D}_{conc}$. An *abstract interpretation* corresponding to this consists of an abstract domain $\langle \mathcal{D}_{abs}, \leq \rangle$ which is also a complete lattice, and a set of "abstract operations" $\widehat{\mathcal{F}}$ over $\mathcal{D}_{abs}$ such that for each concrete operation $f \in \mathcal{F}$ there is an abstract operation $\widehat{f} \in \widehat{\mathcal{F}}$ that "mimics" the execution of f. The abstract and concrete domains are related by a pair of functions

$$abs : \mathcal{D}_{conc} \longrightarrow \mathcal{D}_{abs}$$
$$conc : \mathcal{D}_{abs} \longrightarrow \mathcal{D}_{conc}$$

that are monotone and adjoint, and for each concrete operation f, the corresponding abstract operation $\widehat{f}$ satisfies

$$f(x) \sqsubseteq conc(\widehat{f}(abs(x))) \quad \text{for each } x \in \mathcal{D}_{conc}.$$

---

[14] Recall that the top element of the abstract domain represents all computational possibilities, and conveys no information.

Termination issues are usually addressed by reasoning about the structure of $\mathcal{D}_{abs}$. It suffices to require that $\mathcal{D}_{abs}$ be Noetherian, i.e., contain no infinite ascending chains (though many authors require that $\mathcal{D}_{abs}$ be a finite lattice). If this condition is satisfied, and each abstract operation is monotone, then it is not difficult to see that the least fixpoint of the abstract operations will be computed in a finite number of steps.

The preceding discussion has been couched in operational terms. It is also possible to reformulate this in terms of axiomatic semantics. In this case, $\mathcal{D}_{conc}$ is the lattice of assertions (modulo logical equivalence) ordered by implication, $\mathcal{D}_{abs}$ is a sublattice of $\mathcal{D}_{conc}$, and the meanings of a concrete operation f and its abstract counterpart $\widehat{f}$ are specified as predicate transformers. Now the meaning of a concrete operation can be specified in two ways:

(1) Given a postcondition φ, the meaning of f can be specified in terms of the *weakest precondition* ψ such that $\{\psi\}f\{\varphi\}$.

(2) Given a precondition φ, the meaning of f can be specified in terms of the *strongest postcondition* ψ such that $\{\varphi\}f\{\psi\}$.

Accordingly, any *forward dataflow problem*—that is, one where dataflow information is propagated in the direction of control flow—can be reformulated as an equivalent *backward dataflow problem*—that is, one where dataflow information is propagated in a direction opposite to the direction of control flow. A concrete application of this can be seen in the so called Magic-Set-based approaches to dataflow analysis of logic programs (Codish and Demoen 1993, Codish et al. 1994, Debray and Ramakrishnan 1993, Kanamori 1993, Nilsson 1991), discussed in Section 4.4, where forward dataflow analyses are reformulated as backward analysis problems. A dual approach is taken by Codish et al. (1993a), who reformulate backward analyses in terms of forward analysis problems.

3.3 The widening/narrowing approach

An approach to dataflow analysis that is very different from the Galois connection approach is the *widening/narrowing* approach (Cousot and Cousot 1977). Conceptually, there is no separate domain of descriptions in this case: the analysis proceeds over the concrete domain $\mathcal{D}_{conc}$ (however, the objects considered during analysis must have finite representations to allow the analyzer to manipulate them, and one can consider these finite representations to be "descriptions" in some sense). The basic idea is a straightforward one: first, use a *widening operator* $\triangledown : \mathcal{D}_{conc} \longrightarrow \mathcal{D}_{conc}$ to optain a sound approximation to the meaning of the program; then, use a *narrowing operator* $\triangle : \mathcal{D}_{conc} \longrightarrow \mathcal{D}_{conc}$ to improve this approximation without sacrificing soundness.

Widening: The widening operator $\triangledown : \mathcal{D}_{conc} \longrightarrow \mathcal{D}_{conc}$ is required to satisfy the following conditions:

(1) for every $x, y \in \mathcal{D}_{conc}$, $x \sqsubseteq (x \triangledown y)$ and $y \sqsubseteq (x \triangledown y)$;
(2) for every increasing chain $x_0 \sqsubseteq x_1 \sqsubseteq \cdots \sqsubseteq x_i \sqsubseteq \cdots$ in $\mathcal{D}_{conc}$, the chain $y_0 \sqsubseteq y_1 \sqsubseteq \cdots \sqsubseteq y_i \sqsubseteq \cdots$ defined by

$$y_0 = x_0$$
$$y_{i+1} = y_i \triangledown x_{i+1}, \quad i \geq 0$$

is finite.

Let $\mathcal{X}$ and $\mathcal{Y}$ be the chains $x_0 \sqsubseteq x_1 \sqsubseteq \cdots$ and $y_0 \sqsubseteq y_1 \sqsubseteq \cdots$, respectively, where $y_0 = x_0$; and $y_{i+1} = y_i \triangledown x_{i+1}, i \geq 0$. Then, the first condition implies that $x_i \sqsubseteq y_i$ for all $i \geq 0$, whence it follows that

$$\sqcup \{x_i \mid i \geq 0\} \sqsubseteq \sqcup \{y_i \mid i \geq 0\}$$

i.e., $\sqcup \mathcal{X} \sqsubseteq \sqcup \mathcal{Y}$. In particular, let $x_0 = \bot$ and $x_{i+1} = f^{i+1}(\bot)$ for $i \geq 0$: this is a chain if f is monotone. Then, (1) implies that $\sqcup \mathcal{X} = \mathit{lfp}(f) \sqsubseteq \sqcup \mathcal{Y}$, i.e., $\sqcup \mathcal{Y}$ is a sound approximation to $\mathit{lfp}(f)$, while (2) implies that $\sqcup \mathcal{Y}$ can be computed in a finite number of steps.

Narrowing: The narrowing operator $\triangle : \mathcal{D}_{conc} \longrightarrow \mathcal{D}_{conc}$ is required to satisfy the following conditions:

(1) $(x \triangle y) \sqsubseteq x$;
(2) for every x, y and l in $\mathcal{D}_{conc}$, if $l \sqsubseteq x$ and $l \sqsubseteq y$ then $l \sqsubseteq (x \triangle y)$; and
(3) for every decreasing chain $x_0 \sqsupseteq x_1 \sqsupseteq \cdots \sqsupseteq x_i \sqsupseteq \cdots$ in $\mathcal{D}_{conc}$, the chain $y_0 \sqsupseteq y_1 \sqsupseteq \cdots \sqsupseteq y_i \sqsupseteq \cdots$ defined by

$$y_0 = x_0;$$
$$y_{i+1} = y_i \triangle x_{i+1}$$

is finite.

Note, from the first condition, that $\triangle$ is not commutative. Let $\mathcal{X}$ and $\mathcal{Y}$ be the chains $x_0 \sqsupseteq x_1 \sqsupseteq \cdots$ and $y_0 \sqsupseteq y_1 \sqsupseteq \cdots$, respectively, where

$$y_0 = x_0;$$
$$y_{i+1} = y_i \triangledown x_{i+1}, \quad i > 0.$$

then the first condition implies that $\sqcap \mathcal{X} \sqsubseteq y_{i+1} \sqsubseteq y_i$ for each $i \geq 0$, so that $\sqcap \mathcal{X} \sqsubseteq \sqcap \mathcal{Y} \sqsubseteq x_0$, while the second condition implies that $\sqcap \mathcal{Y}$ can be computed in a finite number of steps. In particular, if for a function $f : \mathcal{D}_{conc} \longrightarrow \mathcal{D}_{conc}$ we have $\mathit{lfp}(f) \sqsubseteq x_0$, and $x_{i+1} = f^{i+1}(x_0)$ for $i \geq 0$, then by induction $\mathit{lfp}(f) \sqsubseteq f^i(x_0)$ for all $i \geq 0$, whence

$$\mathit{lfp}(f) \sqsubseteq \sqcap \mathcal{X} \sqsubseteq \sqcap \mathcal{Y} \sqsubseteq x_0.$$

In other words, $\sqcap \mathcal{Y}$ is a sound approximation to $\mathit{lfp}(f)$ that is at least as precise as x_0, and is—unlike $\sqcap \mathcal{X}$—finitely computable.

Program analysis using widening/narrowing: From the discussion above, it is not difficult to see how widening/narrowing can be used for program analysis. Suppose the meaning of a program is given by a function $f_P : \mathcal{D}_{conc} \longrightarrow \mathcal{D}_{conc}$. The analysis proceeds as follows:

Widening : Compute the limit of the chain $\mathcal{X}$ given by

$$x_0 = \bot$$
$$x_{i+1} = \mathbf{if}\ f_P(x_i) = x_i\ \mathbf{then}\ x_i\ \mathbf{else}\ x_i \triangledown\ f_P(x_i), \qquad i > 0.$$

From the definition of $\triangledown$, $\sqcup \mathcal{X}$ is a sound approximation to $\mathit{lfp}(f_P)$ and can be computed in finitely many steps.

Narrowing : Improve the result obtained from the widening phase by computing the limit of the chain $\mathcal{Y}$ defined as follows:

$$y_0 = \sqcup \mathcal{X};$$
$$y_{i+1} = \mathbf{if}\ f_P(y_i) = y_i\ \mathbf{then}\ y_i\ \mathbf{else}\ y_i \triangle\ f_P(y_i), \qquad i > 0.$$

From the definition of $\triangle$, $\sqcap \mathcal{Y}$ is a sound approximation to $\mathit{lfp}(f_P)$ that is at least as precise as $\sqcup \mathcal{X}$, and is computable in finitely many steps.

3.4 The hybrid approach

Consider again the basic widening/narrowing approach to program analysis discussed above. The central problem in program analysis is that given a program whose meaning is given by a semantic function $f : \mathcal{D}_{conc} \longrightarrow \mathcal{D}_{conc}$, the concrete semantics of the program, given by $\mathit{lfp}(f)$, may take "too long"—i.e., infinitely many steps—to converge. The problem is resolved in the widening/narrowing approach by using the widening operator to "accelerate" convergence to at most finitely many steps, then using the narrowing operator to try and recover some information that may have been discarded by the widening operator.

This idea can be applied to analyses based on the Galois connection approach as well. Of course, if the abstract domain $\mathcal{D}_{abs}$ is Noetherian, the fixpoint computation for the abstract operators will converge in a finite number of steps, so any use of widening/narrowing in this context can be viewed purely as a device for improving the efficiency of the analysis. However, the hybrid approach, combining widening/narrowing with the Galois connection approach, becomes much more significant if $\mathcal{D}_{abs}$ is not Noetherian and contains infinite ascending chains: in this case, the hybrid approach may be terminating where the original analysis, based solely on the Galois connection approach, is not. The reason non-Noetherian abstract domains are interesting is one of precision. Intuitively, the larger the abstract domain, the smaller the "granularity" of abstract domain elements and the "gap" between different abstract domain elements, and hence the more precise the analysis. This suggests that analyses based on abstract domains containing infinite ascending chains may be more precise than Noetherian domains. Indeed, Cousot and Cousot (1991) show that:

(1) For any given program, there is a finite lattice L such that a Galois connection analysis over L gives the same results as widening/narrowing over the concrete domain $\mathcal{D}_{conc}$.

(2) In general, there is no single finite lattice that will do for all programs: if we consider all programs, infinitely many abstract values may be necessary to obtain results equivalent to widening/narrowing over $\mathcal{D}_{conc}$.

(3) For any given program, it is not possible, in general, to infer the set of abstract values needed by a simple inspection of the text of the program.

This suggests that in the interests of precision, it may be necessary to work with non-Noetherian abstract domains, and rely on widening/narrowing to force convergence of computations within a finite number of steps.

How are such "hybrid" widening/narrowing operators to be defined? Cousot and Cousot (1977) give the following guidelines. Suppose there is a Galois connection between $\mathcal{D}_{conc}$ and $\mathcal{D}'_{abs}$, where $\mathcal{D}'_{abs}$ may not be Noetherian, defined by the functions $abs' : \mathcal{D}_{conc} \longrightarrow \mathcal{D}'_{abs}$ and $conc' : \mathcal{D}'_{abs} \longrightarrow \mathcal{D}_{conc}$. Let $\langle \mathcal{D}''_{abs}, \leq \rangle$ be any Noetherian abstract domain such that $abs'' : \mathcal{D}_{conc} \longrightarrow \mathcal{D}''_{abs}$ and $conc'' : \mathcal{D}''_{abs} \longrightarrow \mathcal{D}_{conc}$ define a Galois connection. Then, widening and narrowing operators over $\mathcal{D}'_{abs}$ may be defined as follows:

Widening : Project into $\mathcal{D}'_{abs}$ the join operator $\vee$ on $\mathcal{D}''_{abs}$, as follows: for any $x, y \in \mathcal{D}'_{abs}$,

$x \triangledown y = abs'(conc''(u \vee v))$, where $u = abs''(conc'(x))$, and $v = abs''(conc'(y))$.

Narrowing : Project into $\mathcal{D}'_{abs}$ the meet operator $\wedge$ on $\mathcal{D}''_{abs}$, as follows: for any $x, y \in \mathcal{D}_{abs}'$,

$x \triangle y = abs'(conc''(u \wedge v))$, where $u = abs''(conc'(x))$, and $v = abs''(conc'(y))$.

Examples of such hybrid analyses for logic programming languages include depth abstraction over abstract domain elements (Debray 1992b), and the "star abstraction" of Codish et al. (1991).

3.5 Approximating the greatest fixpoint

The approaches to dataflow analysis described so far have focused on computing an approximation A_0 to the least fixpoint $lfp(f)$ of a concrete semantic function f by computing the least upper bound of an ascending Kleene sequence, starting at $\bot$, such that $lfp(f) \sqsubseteq A_0$.

An alternative approach is to compute an approximation A_1 to the greatest fixpoint of f by computing the greatest lower bound of a descending Kleene sequence starting at $\top$, such that $gfp(f) \sqsubseteq A_1$. Since it is always the case that $lfp(f) \sqsubseteq gfp(f)$, it follows that A_1 is a sound approximation to $lfp(f)$. This is the approach taken, for example, by Codish et al. (1988) and Kifer et al. (1988). In general, the two approaches are not comparable in precision. However, Codish et al. point out a practical advantage of approximating the greatest fixpoint: since the approximation computed by such an analysis is always above $gfp(f)$, it is always a sound approximation to $lfp(f)$. Because of this, it is not necessary to continue the analysis until it converges to a fixpoint: it can be stopped at any point without sacrificing soundness. This approach is used by Taylor in a compiler for FCP (Taylor 1989b): for the particular analysis problem considered, namely, the detection of "input" arguments in FCP programs, Taylor reports that one analysis pass over the program suffices to capture most of the interesting information. A formal treatment of this analysis is given by Codish et al. (1988).

4 Frameworks for dataflow analysis of logic programs

Frameworks for dataflow analysis of logic programs have been proposed by a large number of authors, including Barbuti et. al. (1993), Bruynooghe (1991), Corsini and Filé (1988), Jones and Søndergaard (1987), Kanamori and Kawamura (1987), Marriott and Søndergaard (1988a, 1989), Marriott et al. (1990a), Mellish (1986), Nilsson (1989), and Winsborough (1988). A discussion of each of these frameworks is beyond the scope of this work. Instead, we discuss a few of these frameworks, in order to give the reader a sampling of the wide variety of approaches that have been proposed in this context.

4.1 Frameworks for top-down analyses

4.1.1 *The framework of Bruynooghe*

This is a framework for the top-down analysis of logic programs. The concrete semantics is based on (ordered) AND-trees. In the abstract semantics, control flow is abstracted using abstract AND-OR trees, while sets of substitutions are abstracted using abstract substitutions. To ensure termination, the abstract domain is constrained to have the finite chain property by using *rational* terms, i.e., terms with "back edges", to describe infinite sets of terms. Finally, abstract operations are characterized in terms of four primitive operations on abstract trees: initialization, procedure entry, procedure exit, and abstract interpretation of built-ins.

Concrete AND-trees: Basically, an AND-tree is a proof tree that represents a (partial) SLD-derivation. Each node of the tree is labelled with a pair $\langle A, \sigma \rangle$, where A is a literal and σ a substitution. A computation is represented by a series of such AND-trees. In order to characterize the procedural behavior of logic programs, it suffices to know the instance of each procedure call immediately before it is executed (the "call point") and immediately after it is completed (the "return point"). A sequence of AND-trees representing a computation can therefore be represented using a *generalized AND-tree*, where procedure calls are characterized by their instances at the call and return points. Nodes in a generalized AND-tree can be adorned with substitutions on the left ("call substitutions") and on the right ("return substitutions"); they are represented as $\langle \theta, A: H, \psi \rangle$, where θ is an (optional) call substitution, A is a call, H is a clause head (optional), and ψ a success substitution (optional). Generalized AND-trees are constructed using four operations: *initialization, procedure entry, procedure exit,* and *interpretation of built-ins. Initialization* constructs an initial generalized AND-tree $\langle \theta, A, - \rangle$, given a query for a literal A with call substitution θ. *Procedure entry* and *procedure exit* augment a tree to reflect the effects of a successful resolution step. Finally, *interpretation of built-ins* handles the effects of built-in predicates, primarily unification.

Abstract AND-OR trees: A generalized AND-tree represents a sequence of AND-trees, i.e., an SLD-derivation, or equivalently, a particular execution of a program. For analysis purposes, however, it is necessary to characterize all possible executions of the program, i.e., a set of generalized AND-trees. This can be done using an *abstract AND-OR tree*. In these trees, single substitutions at nodes are replaced by (possibly infinite) sets of substitutions. Given a set of substitutions Θ, a node $\langle \Theta, A, - \rangle$ represents the set of generalized AND-tree nodes $\{\langle \theta, A, - \rangle \mid \theta \in \Theta\}$. Nondeterminism, i.e., the fact that there may be alternative clauses that match a call to a predicate, is handled by introducing OR-nodes. Abstract AND-OR trees are constructed using the following operations: initialization, abstract procedure entry, abstract procedure exit, and abstract interpretation of built-ins. Note that the construction of abstract AND-OR trees becomes nondeterministic, since there is no search rule, and operations are possible in different OR-branches.

Dataflow analysis under this framework then boils down to the following: given a (description of) a set of queries Q, construct an abstract AND-OR tree that is

(1) *sound*, i.e., describes all concrete AND-trees that can occur during the execution of any query in Q;

(2) as precise as possible; and

(3) constructable in a finite amount of time.

The basic approach is to construct abstract AND-OR trees using the four basic operations: initialization, procedure entry, procedure exit, and abstract interpretation of built-ins. The global correctness of the resulting AND-OR tree is ensured by giving sufficient conditions for each of these operations. Intuitively, these represent the correctness requirements for each abstract operation in the Galois connection development of Cousot and Cousot, as expressed in the condition (*) in Section 3.2. To guarantee termination, we need finite representations for (i) infinite sets of substitutions, and (ii) infinite AND-OR trees. Infinite sets of substitutions are handled using *abstract substitutions* that give finite (upper) approximations to infinite sets of substitutions. Suppose the set of concrete and abstract substitutions are denoted by *Subst* and *ASubst* respectively. Then, the set of concrete substitutions Θ that is represented by an abstract substitution S is given by a function

$$\gamma : ASubst \longrightarrow \wp(Subst)$$

such that $\Theta = \gamma(S)$. The union of two sets of concrete substitutions is approximated by an operation $\sqcup : ASubst \times ASubst \longrightarrow ASubst$ such that for any $S_1, S_2 \in ASubst$,

$$\gamma(S_1) \cup \gamma(S_2) \subseteq \gamma(S_1 \sqcup S_2).$$

Infinite AND-OR trees are handled using finite (cyclic) graphs, called *rational trees*, to approximate infinite trees. This approximation can be done in many ways, as long as it is conservative and the resulting set of rational trees satisfies the finite chain property when ordered by inclusion on the set of AND-OR trees they denote.

Examples of analyses based on this framework include integrated type and mode analysis (Bruynooghe and Janssens 1988, Janssens 1991), argument size and termination analysis (Verschaetse and De Schreye 1991), and compile-time garbage collection (Mulkers *et al.* 1994).

4.1.2 *The framework of Marriott, Søndergaard, and Jones*

Marriott, Søndergaard, and Jones describe a somewhat different approach for top-down analysis of logic programs (Marriott *et al.* 1990a) that follows an approach due to F. Nielson. The basic idea is to use a *two-level* description:

(1) Define a meta-language L. Given a language L' of interest, define the (denotational) semantics of L' in terms of L.
(2) Define the semantics of the meta-language L:
   - A "standard interpretation" for L gives the standard semantics for L'.
   - A "non-standard interpretation" for L yields a non-standard semantics for L', and can be used to define an abstract interpretation for L'.

The advantage of such an approach is that the standard semantics, and for each analysis problem the corresponding non-standard semantics for the meta-language together with a proof of soundness, needs to be given only once. These can then be used for a variety of source languages without change.

The meta-language used by Marriott, Søndergaard and Jones is one of typed λ-expressions. The types are given by

$$E \in Exp : S \mid L$$
$$L \in Lat : D \mid E \longrightarrow L$$

where $S \in Stat$ represent *static types*, $D \in Dyn$ represent *dynamic types*, $Stat \cup Dyn$ is the set of *base types*, and Lat is a set of *lattice types*. The difference between static and dynamic types is that the interpretation of a static type is the same in all (standard and non-standard) interpretations, while the interpretation of a dynamic type may change from one interpretation to another. The semantics of this language is given in terms of *type interpretations*. A type interpretation **I** assigns a structure to each base type: a static type is assigned some (fixed) poset, a dynamic type is assigned a complete lattice, and a type $E \longrightarrow L$ is assigned a monotone function from $\mathbf{I}[\![E]\!]$ to $\mathbf{I}[\![L]\!]$. Each lattice type $L \in Lat$ is mapped by **I** to a complete lattice. Marriott, Søndergaard and Jones show that using their framework, the correctness of the entire analysis follows immediately once the base functions have been shown to be safely approximated. Examples of analyses based on this framework include groundness analysis for definite Horn programs (Marriott *et al.* 1990a), analysis of constraint logic programs (Marriott and Søndergaard 1990), and floundering analysis for logic programs containing negation (Marriott *et al.* 1990b).

4.1.3 *The framework of Winsborough: minimal function graphs*

In general, the semantics of a program describes all of its behaviors on all possible inputs. Strictly speaking, therefore, an analysis would have to account for all these behaviors in order to claim that it gave a sound description of the semantics of the program. Often, however, programs are written in order to be executed in certain ways, and it may suffice for an analysis algorithm to capture only these behaviors and ignore others, i.e., to describe only the "relevant portions" of the semantics. For example, consider the following predicate definition:

```
append([], L, L).
append([H|L1], L2, [H|L3]) :- append(L1, L2, L3).
```

Now in principle, any argument of a predicate in a logic program can be used as either an input argument or an output argument, so the predicate append/3 can be used in any of $2^3 = 8$ different ways, and dataflow analyses of this predicate should specify its behavior for all these possibilities. However, suppose we know that in a particular program, this predicate is used in only one way, with the first and second arguments as inputs and the third argument as an output argument. Then, it makes no sense to have an analysis describe all possible behaviors: it suffices to describe only those behaviors of this predicate that pertain to this particular use of the predicate.

This understanding is implicit in much of the work on (top-down) analysis of logic programs (Bruynooghe and Janssens 1988, Bruynooghe *et al.* 1987, Debray 1989, 1992b, Mellish 1986, Muthukumar and Hermenegildo 1991, 1992). An analogous situation arises also in the analysis of functional languages. To formalize this notion, Jones and Mycroft introduced the notion of *minimal function graphs*, which capture only that part of the semantics of a program that is "reachable" from a given set of inputs (Jones and Mycroft 1986). The idea was adapted to logic programming languages by Winsborough (1987, 1988). The idea is to abstract from a *total function graph* semantics that describes all behaviors of a program, and it provides a framework that can be used to formalize a number of top-down analyses of logic programs.

Total function graph semantics: This semantics, which describes the result substitutions that would be produced when a literal in a logic program is executed under each of a set of substitutions, is very similar to various denotational semantics that have been proposed for PROLOG (see, for example, Debray and Mishra 1988, Jones and Mycroft 1984). No information is produced about what procedure calls would be made during a top-down execution. The meaning of a program is a set of monotone mappings that, given a literal and an "input" substitution, produce an "output" substitution.

Minimal function graph semantics: The basic idea here is to define the meaning of a program as a partial function that is defined only on the calls and activation environments that are "reachable" from the input queries assuming some fixed execution strategy, i.e., are actually encountered during a top-down execution of the program. Thus, this assumes that the user provides a set of *entry point specifications*, which are pairs consisting of a literal and an "activation description" over the variables of that literal (intuitively, a description of allowed values for its arguments). Conceptually, we can imagine constructing the entire semantics (as given, for example, by the Total Function Graph semantics) and then restricting it to the "relevant" parts. Of course, more efficient constructions are possible: intuitively, the idea is to construct the minimal function graph semantics in a demand-driven way. For algorithmic aspects of such constructions, the reader may see Debray (1992b); a formal treatment of the semantic issues is given by Winsborough (1988).

4.1.4 *The framework of Kanamori and Kawamura*

The framework of Kanamori and Kawamura (1993), aimed at top-down analyses, is based on OLDT-resolution, which is essentially SLD-resolution augmented with "memoization." The basic idea here is the following: to analyze a PROLOG program, it is necessary to take into account all possible executions. While particular executions can be considered using PROLOG's search strategy (for example, by using a PROLOG interpreter modified to interpret programs over an abstract domain using abstract operations), such a scheme would not, in general, cover all possible executions because PROLOG's search strategy is incomplete. The problem can be rectified by augmenting PROLOG's search strategy with something that will allow all execution paths to be explored—in other words, a top-down search strategy that is complete. Possible approaches include depth-first iterative deepening and SLD-resolution with memoization. The framework of Kanamori and Kawamura is based on the latter.

OLDT-resolution: OLDT-resolution, which augments SLD-resolution with a "memo-table" where calls to a predicate and the solutions returned can be recorded, for possible reuse by later computations, was introduced by Tamaki and Sato (1986), and independently by Dietrich (1987) (who referred to it as "extension tables"), and shown to be complete for Horn logic programs. The basic idea is to maintain, for each predicate p in a program, a table that associates with each call to p a set of answers computed for that call. Execution proceeds as follows: for any call $p(\bar{t})$ to a procedure p, we consult the table for p. There are three possibilities:

(a) If (a variant of) $p(\bar{t})$ is in the table, and is associated with a nonempty set of answers S, then S is returned. The original computation then continues with each element of S.

(b) If (a variant of) $p(\bar{t})$ is in the table but the set of answers is empty, then the call is suspended (or, alternatively, made to fail and eventually recomputed from the beginning).

(c) If no variant of $p(\bar{t})$ can be found in the table, then it is entered into the table associated with the empty set of solutions, after which $p(\bar{t})$ is solved, as follows: for each clause '$H :- B$' defining p:
  * unify the call $p(\bar{t})$ with the head H;
  * if the unification succeeds, process the body goals B using OLDT-resolution;
  * if the body computation succeeds with a substitution θ, then enter $\theta(H)$ into the table as an answer for the call $p(\bar{t})$;
  * "awaken" any suspended computations for $p(\bar{t})$.

Abstract interpretation using OLDT-resolution: The basic idea here is very similar to concrete computation using OLDT-resolution, as described above, the only difference being that abstract operations are used instead of concrete ones. Because OLDT-resolution is a complete evaluation strategy, termination can be guaranteed if the abstract domain is finite. However, analyses may not terminate if the abstract domain is infinite, even if it satisfies the finite chain property. This is illustrated by the following example: consider an analysis whose abstract domain is given by

$$\mathcal{D}_{abs} = \{\bot, \top\} \cup \{\mathbf{list}^i(\mathbf{int}) \mid i \geq 0\}$$

where $\bot$ denotes the empty set, $\top$ the set of all terms, **int** the set of integers, and **list**(x) the set of lists whose elements are in the set denoted by x. $\mathcal{D}_{abs}$ is ordered by $\sqsubseteq$ such that $x \sqsubseteq y$ if and only if the set of elements denoted by x contains the set denoted by y: it is not difficult to see that $\langle \mathcal{D}_{abs}, \sqsubseteq \rangle$ is a flat domain, and obviously satisfies the finite chain property. Now consider the following program:

```
p(X) :- p([X]).
?- p(0).
```

This generates an infinite sequence of mutually incomparable (under $\sqsubseteq$) descriptions of calls:

$p(\mathbf{int}), p(\mathbf{list}(\mathbf{int})), p(\mathbf{list}(\mathbf{list}(\mathbf{int}))), \ldots$

and as a result, the analysis does not terminate.

This problem with termination can be seen in some OLDT-based analysis algorithms for logic programs, e.g., that of Bansal and Sterling (1988), which is not guaranteed to terminate on all inputs. It can be handled using depth abstraction, as suggested by Debray (1992b), or other schemes conceptually related to the "widening" operation of Cousot and Cousot (1977).

Examples of analyses that are based on, or fit into, this framework include mode analysis (Debray 1989, Debray and Warren 1988), success pattern analysis (Kanamori and Kawamura 1987), type inference (Horiuchi and Kanamori 1987, Kanamori and Horiuchi 1985), functionality analysis (Kanamori et al. 1987a), and termination analysis (Kanamori et al. 1987b).

4.2 Frameworks for bottom-up analyses

Frameworks for bottom-up analyses are based on various model-theoretic semantics for logic programs. Here we discuss two such frameworks: one due to Marriott and Søndergaard (1988a, 1992), and one due to Barbuti et al. (1993). Other bottom-up analysis frameworks that have been dsecribed in the literature include that of Kemp and Ringwood (1990), which is discussed in Section 4.3.

4.2.1 The framework of Marriott and Søndergaard

The framework of Marriott and Søndergaard (1988a, 1992), is intented for the analysis of normal logic programs. It is interesting in that it was the first proposal for the analysis of logic programs based on the model-theoretic semantics, even though such semantics had been proposed over a decade earlier (van Emden and Kowalski 1976), and among the relatively few semantics-based approaches to the analysis of logic programs that addresses the treatment of negation.

The collecting semantics used in the analysis derives from the three-valued semantics of Fitting (1985). Intuitively, an interpretation for a program is a pair $\langle u_s, u_f \rangle$ where u_s specifies the set of (ground) atoms that are true, while u_f specifies those that are false. The collecting semantics captures information about which ground atoms are true and which are false, and also which ground instances of program clauses have bodies that succeed and which ground atoms have finitely failed proof trees (note that for failure information, it suffices to collect the heads of ground instances of clauses that have failed: it is unnecessary to collect the bodies of such clauses).

Among the applications considered by the authors are type inference and termination analysis.

4.2.2 The framework of Barbuti et al.

This framework, discussed by Barbuti et al. (1993), is aimed at bottom-up dataflow analyses of logic programs. It is based on the generalized model-theoretic semantics of Falaschi et al. (1989), which defines the meaning of a program in terms of sets of (possibly non-ground) atoms constructed using the function and predicate symbols appearing in the program. The framework is based on two levels: the *core level* defines an abstract T_P operator, which specifies the meaning of a program P, in terms of a set of primitive operators, while the *specialization level* defines the primitive operators for any particular analysis. The essential idea is to define the least algebraic structure that supports the notions of composition (i.e., "collecting together" of results) and interpretations.

The abstract and concrete domains: Given a program P, let B_P denote the set of all (ground and non-ground) atoms that can be constructed using the predicate and function symbols of P. The *generalized Herbrand base* $\mathcal{H}_P$ is the lattice $\langle \wp(B_P), \subseteq \rangle$. For the purposes of analysis, an *abstract Herbrand base* $\langle \mathcal{H}_P^\#, \sqsubseteq \rangle$ can be derived as a homomorphic image of $\langle \mathcal{H}_P, \subseteq \rangle$, where the homomorphism specifies which concrete values are to be considered to be "equivalent" for analysis purposes. $\langle \mathcal{H}_P^\#, \sqsubseteq \rangle$ is a complete lattice, and $\sqsubseteq$ captures the notion of an element of the abstract Herbrand base containing more information than another: recall, from the discussion in Section 3.1, that for analysis purposes we are concerned with precision, and in this context the higher an element is in the concrete or abstract Herbrand base, the less precise it is, and therefore the less information it contains. From this perspective, if $I_1 \sqsubseteq I_2$ then I_1 contains more information than I_2.

Abstract substitutions: To capture some observable program properties, it is necessary to model the notion of substitutions over the abstract domain. Again, some distinct concrete values—in this case, terms—will not be distinguished for the purposes of any particular analysis. This gives rise to a set of *abstract terms*, which is a homomorphic image of the extended Herbrand universe U_P of the program P, i.e., the set of (ground and non-ground) terms that can be constructed from the function symbols and constants appearing in the program. An *abstract substitution* is a mapping from a finite set of variables to the set of abstract terms. For program analysis purposes, it suffices to restrict our attention to abstract substitutions that are defined only on finite set of variables appearing in the program being analyzed. Finally, properties of interest for a particular analysis can be captured via an equivalence relation $\approx_S$ on abstract substitutions. The poset of abstract substitutions over a set of variables V is written $\langle \Sigma_V^\#, \leq_\Psi \rangle$.

An abstract interpretation can then be defined as follows:

Concrete domain : This contains two components: the generalized Herbrand base $\langle \wp(B_P), \subseteq \rangle$, and the set of sets of substitutions over the program variables V, written $\langle \wp(Subst_V), \subseteq \rangle$.

Abstract domain : By analogy with the concrete domain, this also contains two components, the abstract Herbrand base $\langle \mathcal{H}_P^\#, \sqsubseteq \rangle$ and the set of abstract substititions $\langle \Sigma_V^\#, \leq_\Psi \rangle$ over the program variables V.

Galois connections : Relationships between the abstract and concrete domains are established via two Galois connections, one between the lattices $\langle \wp(B_P), \subseteq \rangle$ and $\langle \mathcal{H}_P^\#, \sqsubseteq \rangle$, the other between $\langle \wp(Subst_V), \subseteq \rangle$ and $\langle \Sigma_V^\#, \leq_\Psi \rangle$.

Abstract operations : There are three abstract operations: abstract unification (written $\alpha\text{-}mgu$); abstract union, which is the join operator $\sqcup$ on $\mathcal{H}_P^\#$; and abstract substitution application (written $\alpha\text{-}apply$). These operations are required to satisfy the local correctness conditions (*) for correctness of the entire analysis.

The abstract fixpoint semantics of a program is then defined as follows: first, for any clause $C \equiv \text{'}A :- B_1, \ldots, B_n\text{'}$ in the program, use $\alpha\text{-}apply$ and $\alpha\text{-}mgu$ to define an operator $T_C^\# : \mathcal{H}_P^\# \longrightarrow \mathcal{H}_P^\#$ as follows:

$$T_C^\#(I) = \alpha\text{-}apply(A, \alpha\text{-}mgu(B_1, \ldots, B_n), (B_1', \ldots, B_n'))$$
where $\{B_1', \ldots, B_n'\} \subseteq I$.

Then, given a program defined by clauses $\{C_1, \ldots, C_m\}$, the *abstract immediate consequence operator* $T_P^\# : \mathcal{H}_P^\# \longrightarrow \mathcal{H}_P^\#$ for the program is given by $T_P^\#(I) = \sqcup_{i=1}^m T_{C_i}^\#(I)$. The abstract fixpoint semantics of a program P is given by $\mathcal{F}^\#(P) = lfp(T_P^\#)$.

It can be shown that if the abstract operators $\alpha\text{-}mgu$, $\sqcup$ and $\alpha\text{-}apply$ satisfy the local correctness condition (*), then $\mathcal{F}^\#(P)$ is a sound description of the concrete fixpoint semantics of the program P, i.e., the analysis is correct. Termination is addressed by assuming that $\langle \mathcal{H}_P^\#, \sqsubseteq \rangle$ is a finite lattice.

Analyses based on this framework include type inference (Barbuti and Giacobazzi 1992), data dependence analysis (Giacobazzi and Ricci 1990), and determinacy analysis (Giacobazzi and Ricci 1992).

4.3 Algebraic approaches

Kemp and Ringwood: Kemp and Ringwood (1990) describe a framework for abstract interpretation of logic programs that is purely algebraic, and does not need to concern itself with linguistic details such as substitutions. Their approach is based on the generalized model-theoretic semantics of Falaschi et al. (1989). The basic idea is to construct the abstract domain using algebraic operations on subsets of the concrete domain instead of using "abstract substitutions." They show that by appropriate choice of the concrete domain, it is possible to describe both top-down and bottom-up analyses in their framework.

Let E be the least set containing variables and the atomic formulae of the language under consideration and closed under finite tupling. The concrete domain $\mathcal{D}_{conc}$ is the quotient of E, augmented with a greatest element $\top$, with respect to variable renaming. Let $\sqsubseteq$ be the partial order "less general than" over terms, extended in the natural way to atoms and tuples, then $\langle \mathcal{D}_{conc}, \sqsubseteq \rangle$ is a complete lattice.

The abstract domain is constructed starting with a collection S_0 of subsets of $\mathcal{D}_{conc}$ that contains $\emptyset$ and $\cup S_0$, extending the join operator $\sqcup$ ("most general instance") to S_0 elementwise in the natural way (ensuring that the result is a join operation requires "closing up" the elements of S_0 so that $\sqcup$ is idempotent), and finally closing the resulting sets under intersection.

The concrete domain operations $\cup$ (which captures "merging" the results from different computational paths) and $\sqcup$ (which captures the effects of unification) are approximated in the abstract domain via the operations $\cup^{\#}$ and $\sqcup^{\#}$ respectively, defined as follows:

$$A_1 \cup^{\#} A_2 = \cap \{A \in \mathcal{A} \mid (A_1 \cup A_2) \subseteq A\}; \text{ and}$$
$$A_1 \sqcup^{\#} A_2 = \cap \{A \in \mathcal{A} \mid (A_1 \sqcup A_2) \subseteq A\}$$

Because syntactic objects such as goals, clauses, proof trees, etc., are treated uniformly in this framework using (nested) tuples of objects, these two operations suffice to describe different dataflow analyses.

Giacobazzi et al.: A somewhat algebraic framework for program analysis is described by Giacobazzi et al. (1992). The intent here is to provide an algebraic framework that is general enough to account for analyses of constraint logic programs, of which ordinary logic programs can be seen as a special case. To this end, the authors begin with a general notion of "term systems," which intuitively capture (via appropriate axioms) the notions of variables, terms, and substitutions, and then introduce operations and axioms for conjunction, disjunction, existential quantification, and equality. The resulting algebraic structures are essentially *closed semirings* (Aho et al. 1974), enriched with cylindrification operators (to deal with existential quantification) and "diagonal elements" (to deal with equality constraints). The idea is to define the standard as well as various non-standard semantics by appropriately choosing the underlying sets of terms, primitive constraints, and operations.

4.4 "Magic-Set"-based approaches

As discussed in Section 3.2, every forward dataflow analysis can also be formulated as a backward analysis, and vice versa. Since top-down analyses of logic programs are essentially forward analyses while bottom-up analyses are backward analyses, this indicates that top-down analyses can be formulated equivalently as bottom-up analyses. Techniques based on the *magic sets transformation*, originally devised for the efficient evaluation of queries in deductive databases (Bancilhon *et al.* 1986) offer one way to accomplish this without any (asymptotic) loss in efficiency (Ramakrishnan and Sudarshan 1991, Ullman 1989). They also provide a uniform treatment of both top-down and bottom-up analyses in terms of the model-theoretic semantics of logic programs, which are usually much simpler than the various top-down denotational or operational semantics that have been proposed. Dataflow analysis of logic programs using magic sets techniques have been proposed by Codish and Demoen (1993), Codish *et al.* (1994), Debray and Ramakrishnan (1993), Kanamori 1993, Mellish (1990) and Nilsson (1991).

The "Magic Sets" transformation: This involves rewriting a given program in such a way that the fixpoint evaluation of the rewritten program is efficient, in that facts that are not relevant to answering a given query are not generated. The rewriting essentially modifies the rules in the original program by adding literals that act as "filters," preventing the generation of irrelevant facts. Further, new rules defining the predicates in these literals are added to the program. These predicates in effect compute the set of goals that are invoked a top-down (PROLOG-style) evaluation of the original program.

The rewriting proceeds in two phases. Given a program and a query, the first phase of the algorithm processes the clauses of the program to determine an order of evaluation for the body literals: in the process, it is able to determine which arguments of a given body literal will become bound by the time that literal is executed. Literals are annotated with information about bound and free arguments (this information is used to then determine an ordering for the body literals of the clauses for that literal), producing what is called an *adorned program*. In the second phase, the final program is generated by a rewriting algorithm from the adorned program using information about the ordering of body literals.

The first phase of the algorithm, whose most important aspect is the decision about the order in which the body literals in a clause are to be processed, is not directly relevant to the application of magic sets techniques to program analysis, since for the purposes of dataflow analysis we assume that we are given the control strategy of the program (which dictates the ordering on body literals).

In the second phase, the rewriting proceeds by creating a new predicate $magic\_p(\bar{x})$ for each predicate p in the program.[15] Each clause in the program is augmented with a call to such a "magic" predicate that acts as a filter that prunes away computations that are provably irrelevant to the top-level query under consideration, and new clauses are added to the program to define these "magic" predicates. As an example, consider an adorned clause

$$q_0(\bar{x}_0, \bar{y}_0) :- q_1(\bar{x}_1, \bar{y}_1), \ldots, q_n(\bar{x}_n, \bar{y}_n)$$

where the arguments $\bar{x}_i$ represent the bound, or "input," arguments, while the $\bar{y}_i$ represent the unbound, or "output" arguments, $0 \leq i \leq n$. Suppose that the body literals are executed in their left-to-right order. Then, this clause is transformed to

$$q_0(\bar{x}_0, \bar{y}_0) :- magic\_q_0(\bar{x}_0), q_1(\bar{x}_1, \bar{y}_1), \ldots, q_n(\bar{x}_n, \bar{y}_n).$$

In addition, the following clauses defining the "magic" predicates $magic\_q_i$, corresponding to the body literals q_i, $1 \leq i \leq n$, are added to the program:

$$magic\_q_i(\bar{x}_i) :- magic\_q_0(\bar{x}_0), q_1(\bar{x}_1, \bar{y}_1), \ldots, q_{i-1}(\bar{x}_{i-1}, \bar{y}_{i-1}).$$

Finally, a "seed predicate" $magic\_p(\bar{x})$ is added corresponding to the top-level query $p(\bar{x}, \bar{y})$, where $\bar{x}$ represents the bound arguments of the query.

Program analysis using the "Magic Sets" transformation: Analysis of logic programs using the magic sets transformation involves three steps. The program is first rewritten to make the primitive operations such as unification and application of substitutions explicit. The primitive operations are then replaced by the corresponding "abstract operations" to obtain an abstract program. This program is then transformed using the magic sets rewriting algorithm described above. The resulting program can then be evaluated bottom-up using standard techniques: when the least fixpoint is computed, for each predicate q in the program the facts computed for $magic\_q$ give the calling patterns for q that can arise when the program is executed, while the facts for q give the success patterns for q.

If the transformation is carried out in exactly this manner, the connection between calling and success patterns is lost, and this can sometimes lead to a loss in precision (Debray and Ramakrishnan 1993). Nilsson shows that such an analysis is essentially identical to Mellish's original algorithm for mode inference (Mellish 1981). This loss in precision can be avoided

---

[15]The adornments of literals, i.e., encoding of bound and free arguments, also play a role in this transformation—these details are not directly relevant to our purposes, and have been omitted for simplicity.

by explicitly adding "calling" and "return" argument tuples to the predicates before transforming the program, so that the connection between the calling and success patterns is not lost (Debray and Ramakrishnan 1993).

5 Complexity and efficiency issues

It is well-known that nontrivial program properties are undecidable (cf. Rice's theorem in recursion theory), so the best that dataflow analyses can hope to achieve is to get (conservative) approximations to such properties. Intuitively, it would seem that the more work an analysis algorithm does, the more information it will be able to collect, and therefore the more precise it will be. In this section, we discuss this relationship between efficiency and precision of dataflow analysis algorithms from both theoretical and practical viewpoints.

5.1 The relationship between complexity and precision

To quantify the theoretical relationship between the precision of an analysis algorithm and its complexity, it is necessary to define the notion of "precision" in a precise way. A natural characterization of the precision of a flow analysis algorithm is relative to some other algorithm: the first is more precise than the second if and only if for every program and every input, the results obtained using the first algorithm are uniformly more precise than those obtained from the second. Cousot and Cousot use such a characterization to show that the space of all abstract interpretations of any given language for a given abstract domain, when ordered according to precision, form a complete lattice, and thereby infer the existence of a "most precise" abstract interpretation (Cousot and Cousot 1977, 1979). Because this expresses the precision of an analysis in terms of that of other analysis algorithms, we refer to this notion of precision as *relative precision*.

However, it is not immediately obvious how such a notion can be used to obtain insights into the intrinsic computational costs associated with analyses of different degrees of precision. An alternative approach, discussed by Debray (1992a), considers the precision of an analysis to be given by the class of programs for which the analysis is exact, i.e., the class of programs for which the concretization of the fixpoint computed by the analysis coincides with the collecting semantics of the program. We refer to this notion of precision as *absolute precision*. It can be shown that for any two analysis algorithms A and B, if A is relatively more precise than B then it is also absolutely more precise than B, but the converse need not hold in general.

Using the notion of absolute precision, the following complexity results are reported by Debray (1992a):

(1) Precise analysis of programs that contain no aliasing, where there are only two distinct constants and no function symbols of nonzero arity, is EXPTIME-complete.

The problem is PSPACE-hard under the additional constraint that there is no recursion.

The problem is NP-complete if, in addition to the constraints listed above, we require also that the maximum arity of any predicate symbol be $O(1)$; it remains NP-complete even if the maximum arity is restricted to 3.

(2) If we allow function symbols of nonzero arity, then precise analysis of programs that contain negation, but contain no recursion or aliasing, is NEXPTIME-hard. It remains NEXPTIME-hard even if there are no function symbols with arity greater than 0. This addresses the complexity of analyses that attempt to treat negation in a precise way, e.g., those of Marriott and Søndergaard (1988a, 1992).

The problem is PSPACE-hard for programs contain no negation, recursion or aliasing, and satisfy the additional constraint that the maximum arity of predicate and function symbols be $O(1)$. It remains PSPACE-hard even if the maximum arity of any function or predicate symbol is restricted to 2, and the maximum number of literals in the body of any clause is restricted to 2.

(3) Precise groundness and alias analyses of programs that contain no function symbols, where there are no failed execution branches at runtime, is EXPTIME-complete. This addresses one aspect of the worst-case complexity of a groundness analysis proposed by Marriott et al. (1990a).

The problem is co-NP-complete if, in addition to the above constraints, we require also that there be no recursion, and that the maximum arity of any predicate be $O(1)$; it remains co-NP-complete even if the maximum arity of any predicate is restricted to 6.

The problem remains co-NP-complete even if each clause is allowed to have at most one success pattern: this addresses the worst-case complexity of any algorithm whose precision is similar to that of a groundness analysis proposed by Codish et al. (1994).

(4) Analysis algorithms with polynomial-time worst case complexities can be obtained if dependencies between variables can be ignored or if such dependencies can be assumed to be transitive, and if the number of distinct calling and success patterns for any predicate is $O(1)$ (Debray 1989, 1992b).

5.2 Deriving efficient analysis algorithms

The results quoted in the previous section make it unlikely that we will be able to find analysis algorithms that are both efficient and precise for all programs. The best that we can hope for, instead, are algorithms that are "reasonably precise" for "reasonably large" classes of programs. Here we discuss some approaches that have been taken towards this end. We consider three main approaches: the first focuses on properties of the abstract domain to obtain efficient algorithms; the second relies on carefully engineered algorithms that perform well for "common" programs; and the third uses ideas from partial evaluation to reduce the cost of interpreting programs over the abstract domain. Examples of the first include the work of Debray (1992b), which relies on closure properties of abstract domain elements to safely ignore aliasing effects, yielding analysis algorithms that are efficient for all programs, but where there may be a loss of precision in some cases; and that of Jacobs and Langen (1989) and Langen (1991), which relies on the structure of the abstract domain to simplify fixpoint computations during analysis. An example of the second class is the work of Le Charlier *et al.* (see Englebert *et al.* 1993, Le Charlier *et al.* 1991, Le Charlier and Van Hentenryck 1994), which considers a generic abstract interpretation algorithm and discusses various optimizations as well as its complexity (as measured in the number of iterations needed to compute a fixpoint) for various common classes of programs. Examples of the third approach are discussed by Hermenegildo *et al.* (1992).

Ignoring aliasing effects This approach is motivated by the observation that handling aliasing effects in a precise way can be a source of significant source of overhead. Since many analyses do not really need aliasing information, and most "commonly encountered" programs do not exhibit a great deal of aliasing, this suggests that ignoring aliasing effects may yield efficient analysis algorithms. The central question that arises, if aliasing effects are to be ignored, is: when can this be done without compromising soundness? The answer to this question is intuitively fairly obvious: if aliasing effects are ignored during an analysis, then the binding of a variable that is aliased may change from a value v to another value v', where v' is a substitution instance of v, without the analysis being aware of this. To ensure that the analysis is sound, then, we have to ensure that every element of the abstract domain is closed under substitution. It can be shown that a substitution-closed abstract domain can be constructed from any finite set of sets of terms, essentially by first adding top and bottom elements, then closing each element under substitutions, and finally by closing the resulting set under intersection. This yields a finite set of sets of terms that forms a Moore family, and forms the basis for the dataflow analysis algorithm described by Debray (1992b).

The algorithm described by Debray (1992b) is based on OLDT-resolution. Its worst case complexity is parameterized with respect to the cost of deciding the equality of any two arbitrary elements of the abstract domain. If this cost is C, then assuming that the maximum arity of any predicate and function symbol in the program is $O(1)$ and the number of calling and success patterns for any predicate is $O(1)$, the complexity of the algorithm for a program of size n is $O(nC)$. The justification behind these assumptions is that first, the arity of predicate and function symbols typically does not increase as the size of a program increases—rather, an increase in the size of a program is typically because the number of clauses in it has increased; and secondly, practical experience suggests that in most cases, a procedure in a program is called in only a very few different ways, so that most of the various different ways in which it could be called are never encountered.

Reducing the cost of fixpoint computation A very different idea for reducing the cost of dataflow analysis of logic programs is given by Jacobs and Langen (1989), who propose a scheme to preprocess the clauses defining each predicate to construct a "summary" of its possible behaviors, which can then be used to efficiently construct the actual behavior for any particular call to that predicate. The analysis described by Jacobs and Langen (1989) focuses on groundness of variables and aliasing between them, the intended application being parallelizing compilers for PROLOG. The first step is to "preanalyze" each predicate definition to generate a fixed approximation to its meaning. This step, which the authors call *condensing*, may require iterating to a fixpoint. After this, the effects of any call to a predicate can be analyzed by a single "abstract unification" step between the call and the fixed approximation computed during condensing. The motivation behind this approach is that a predicate is defined once but called from many places, so it may be more efficient to iteratively compute the abstract meaning of the predicate once, then using a single abstract unification step to compute the behavior of each call to that predicate.

The key step in this approach is the condensing step. Jacobs and Langen show that there is no loss of precision during condensing provided that the abstract operations satisfy certain algebraic properties:

(1) abstract unification is "associative", so that the order in which a set of operations is performed does not affect performance; and
(2) abstract unification is additive, so that "least upper bound" operations do not lose information.

The particular analysis considered by the authors satisfy these properties: for example, the first property is satisfied because grounding a variable (i.e., binding it to a ground term) before aliasing is equivalent, in their abstract domain, to aliasing the variable and then grounding it.

The authors use a denotational semantics for (pure) PROLOG as their concrete semantics, and the abstract meaning of a predicate is a function from a description of the input(s) to a description of the output(s). The overall analysis algorithm proceeds as follows: first, for each predicate in the program, the abstract meaning is computed for the most general call, i.e., where each argument is a distinct variable, using condensation. After this, the meaning of any particular call to a predicate can be obtained by a single abstract unification step using the precomputed meaning, without having to iteratively compute a fixpoint.

OLDT-based analysis vs. condensing: OLDT-resolution and condensing are two alternative fixpoint computation strategies for a "generic" abstract interpretation procedure. The main difference between the two approaches is that condensing involves a single fixpoint computation procedure and multiple abstract unification steps, while OLDT-resolution may require multiple fixpoint computations. To the best of our knowledge, there have been no careful empirical studies to determine which performs better in practice. However, based on the differences between the two algorithms, we can draw some broad conclusions.

Since the savings in condensing come from avoiding repeated fixpoint computations, it would seem that this approach would do better than OLDT-based analyses if the predicates in a program were called in many different ways. On the other hand, if the predicates are called in only a small number of different ways, then it may happen that the work invested in the condensing approach in computing the most general meaning of each predicate is not compensated for by the savings obtained from avoiding subsequent fixpoint computations. This suggests that OLDT-based analyses may perform better if predicates are called in a small number of different ways, while condensing may be more efficient if they are called in a large number of different ways. Both are equally efficient if predicates are always used in their "most general" modes, i.e., with all arguments being distinct variables.

Experience suggests that in practice, programmers write programs in such a way that predicates are usually called in a rather small number of different ways. This suggests that in practice, OLDT-based analyses may be more efficient than condensing-based ones.

Efficient generic dataflow analysis algorithms: The approaches of Debray (1992b) and Jacobs and Langen (1989) described above are aimed at exploiting properties of the abstract domain to obtain efficient analysis algorithms. An alternative is to consider a "generic" analysis algorithm (i.e., where the abstract domain, as well as operations that depend on the abstract domain, are considered to be parameters to the algorithm) and optimize it carefully. This is the approach taken by Le Charlier *et al.* (see Englebert *et al.* 1993, Le Charlier *et al.* 1991, Le Charlier and Van Hentenryck 1992a, 1994). The overall structure of the algorithm is similar to other dataflow analyses: namely, the behavior of a program is described by a set of (recursive) equations, which are then solved by iteratively computing a least fixpoint. The main innovation in the approach of Le Charlier *et al.* is the observation that any particular dataflow analysis algorithm must specify how certain decisions are to be made, e.g., when and how to test for termination, how to choose the "next" execution branch to examine at any point, etc., and that careful choices here can reduce the computational cost of the algorithm regardless of the abstract domain under consideration. The authors also identify some other sources of optimization that do not depend on any particular abstract domain. The result is a "generic" abstract interpretation algorithm that has been carefully optimized to avoid unnecessary work wherever possible.

A detailed discussion of the optimizations considered by Le Charlier et al. is beyond the scope of this paper. Here we give some examples of how their algorithm avoids unnecessary work, to give a flavor of the kinds of optimizations they carry out. One optimization involves termination testing. Instead of explicitly comparing the old and new values computed during each iteration of the analysis, it is possible to maintain a Boolean flag that is set if any change is made at any time: after each iteration, this flag can be checked to determine whether the analysis should terminate. This is a fairly obvious optimization that results in some savings by avoiding a potentially nontrivial comparison to check for termination. The authors observe, however, that this opens up further avenues for optimization: since the old value during an analysis no longer needs to be maintained separately, it can be updated destructively during analysis, resulting in significant savings. Another optimization they consider involves *definitive values*, i.e., abstract values that will not change at any later point during analysis. The idea is to keep track of whether an abstract value can change later in the analysis due to computations that are currently suspended (see the discussion on OLDT-resolution in Section 4.1.4): if there are no such suspended computations, this value will not change later in the analysis, and need not be reconsidered from then on.

Worst-case complexity analyses for such a generic algorithm, for different classes of programs, is given by Le Charlier et al. (1991). Let h denote the height of the abstract domain, s the size of the abstract domain, and N the size of the program (in terms of the number of literals). The authors report the following results:

- For programs containing no recursion, the algorithm takes $O(N)$ iterations in the worst case.
- For programs containing only direct recursion, the algorithm takes $O(Nh)$ iterations for a class of programs called "substitution-preserving", and $O(Nhs^2)$ iterations for general direct-recursive programs.
- For programs involving mutual recursion, if we assume that each mutually recursive clause calls only base predicates and one other mutually recursive predicate, then the number of iterations is $O(Nhm)$, where m is the maximum number of predicates in a mutually recursive cycle.
- For arbitrary logic programs, the algorithm takes $O(N^2hs^2)$ itera-

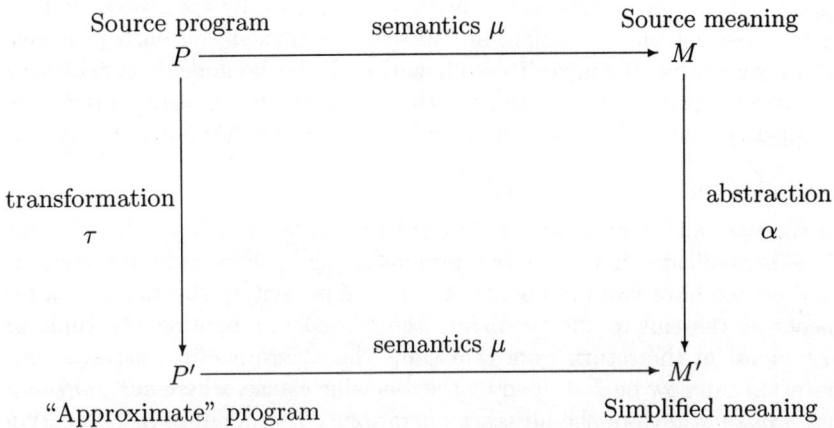

FIG. 2. Analysis, abstraction and "approximate" programs.

tions in the worst case.

It should be noted, however, that on the one hand, the results of Section 5.1 indicate that the worst case complexity of an analysis can be exponential in the size of a program even if the program is non-recursive, and therefore is analyzable, in principle, in one iteration; on the other hand, empirical results for imperative languages indicate that most programs encountered in practice require at most 4 or 5 iterations to reach a fixpoint (Aho et al. 1986), and it seems plausible that similar results may apply to logic programs as well.

Partial evaluation: A straightforward implementation of a global flow analysis system, based on the technique suggested by the name "abstract *interpretation*," might proceed by modifying a standard meta-circular interpreter to compute over the abstract domain—indeed, this is the approach taken in the implementations described by Le Charlier and Van Hentenryck (1994) and Van Roy (1990). An alternative is to specialize such an abstract interpreter to deal with only the program under consideration. This can be done by making a single pass over the program P to be analyzed and producing a transformed program $P' = \tau(P)$ which, when executed, yields precisely the desired flow information about the original program P (see Figure 2). This transformation can be thought of as a partial evaluation of the abstract interpreter with respect to the input program P being analyzed (Codish 1986).

The transformation τ is determined by the flow information desired. Abstract interpretation of a program consists essentially of "simulating" its execution over an abstract domain. This is done by specifying, as part of the abstract interpretation, an "abstract operation" for each primitive operation of the language. To see how this should be done, it is necessary to make the primitive operations of the language – in our case, application of substitutions and unification – explicit. Consider the clause

$$p(\bar{T}_0) :- q_1(\bar{T}_1), \ldots, q_n(\bar{T}_n).$$

In the specialized program, each k-ary predicate—which can be thought of as a predicate that takes one argument that is a k-tuple of terms—is modified to have two arguments: the first representing the tuple of arguments at the call to the predicate, the second representing the tuple of arguments at the return from that call. The corresponding abstract computation can now be described by the following clause, where *abs_unify* and *abs_app_subst* denote the abstract operations corresponding to unification and application of substitutions:

$$\begin{aligned}
abs\_p(\bar{X}_{in}, \bar{X}_{out}) :- \\
& abs\_unify(\mathbf{Id}, \bar{X}_{in}, \bar{T}_0, A_0), \\
& abs\_app\_subst(A_0, \bar{T}_1, \bar{T}_{1,in}), \\
& abs\_q_1(\bar{T}_{1,in}, \bar{T}_{1,out}), \\
& abs\_unify(A_0, \bar{T}_{1,in}, \bar{T}_{1,out}, A_1), \\
& \ldots, \\
& abs\_app\_subst(A_{n-1}, \bar{T}_n, \bar{T}_{n,in}), \\
& abs\_q_n(\bar{T}_{n,in}, \bar{T}_{n,out}), \\
& abs\_unify(A_{n-1}, \bar{T}_{n,in}, \bar{T}_{n,out}, A_n), \\
& abs\_app\_subst(A_n, \bar{T}_0, \bar{X}_{out}).
\end{aligned}$$

Here **Id** represents the abstract domain element corresponding to (the singleton set containing) the identity substitution. The A_i are "abstract substitutions", i.e. abstract domain elements representing sets of substitutions. The resulting program is referred to as the "approximate" program. In general, such a program has to be executed using a complete evaluation strategy, e.g., OLDT-resolution, to ensure that all execution paths are explored and that execution terminates. The program so produced can be executed using an ordinary PROLOG system, provided that the OLDT-mechanism (or whatever complete evaluation strategy is being used) is "folded into" the program: for a more complete discussion, see Hermenegildo *et al.* (1992).

The practical benefit of this approach is that since the flow information is obtained by executing the transformed program directly, instead of having the underlying system execute the abstract interpreter which in turn symbolically executes the original program, one level of interpretation is avoided during the iterative fixpoint computation characteristic of dataflow analyses.

5.3 Experimental results

In this section, we discuss results from three experimental studies of dataflow analysis systems for PROLOG programs. The systems described all implement top-down analysis algorithms.

Hermenegildo, Warren, and Debray: An early study by Hermenegildo et al. (1992) tested two different dataflow analysis systems—namely, MA^3, an analysis system intended for use in an AND-parallel PROLOG system; and Ms, an experimental mode inference system for SB-PROLOG—on a variety of small and medium-sized programs. Both these systems were OLDT-based systems written in PROLOG, and used program transformation to improve the speed of analysis. The broad conclusions of this work were that (i) dataflow analysis accounted for about 30% to 80% of the total compilation time, i.e., increased total compile time by a factor of about 3 to 5; and (ii) with regard to precision, the percentage of "hits", i.e., argument positions about which the analyses produced accurate information, ranged from 35% to 100%, with the exception of one benchmark, on which precision dropped to 11.5%, primarily because the analyzers did not have some relatively sophisticated information about the builtins functor/3 and arg/3.

Van Roy: Van Roy describes a dataflow analyzer implemented as part of the Aquarius PROLOG compiler (Van Roy 1990). The entire compiler, including the dataflow analysis component, is implemented in PROLOG. The analysis is intended to provide information for low-level compile-time optimizations, e.g., whether a variable is ground, uninitialized, dereferenced, etc. It uses two heuristics to balance the tradeoff between efficiency and precision: first, it does not traverse a clause if that clause is already "active", i.e., being analyzed (this is similar to the treatment of recursive clauses in analyses based on OLDT-resolution); and second, a predicate is analyzed if and only if two conditions hold: (i) the entry type of the predicate has changed since it was last analyzed; and (ii) at least one of the clauses for that predicate is active.

For a representative set of realistic PROLOG benchmarks, the analysis is able to derive type information for 56% of the predicate argument positions. For most programs, the analysis time is found to be roughly proportional to the number of argument positions for the predicates.

Le Charlier and Van Hentenryck: Le Charlier and Van Hentenryck have carried out extensive experimental work on abstract interpretation of logic programs (see Le Charlier et al. 1991, Le Charlier and Van Hentenryck 1992a, 1992b, 1994). Le Charlier and Van Hentenryck (1994) describes experimental studies of the generic abstract interpretation algorithm discussed by Le Charlier et al. (1991). The implementation is written in PASCAL: this makes the execution overhead smaller than the systems described in Hermenegildo et al. (1992); however, partial evaluation of the abstract interpreter is not as straightforward. For the purposes of experimentation, the authors consider an 8-element abstract domain for groundness analysis. The broad conclusions of this work are (i) the cost, measured by the number of iterations needed to compute a fixpoint, typically varies linearly with program size, averaging about $3n$ for a program of size n; (ii) the cost, measured by the execution time, typically also varies linearly with program size, averaging about $0.6n$ for a program of size n; and (iii) the precision of analysis, as measured by the number of "hits", range from 76% to 100%.

Le Charlier and Van Hentenryck (1992a) describes experimental results for *reexecution*, a strategy for improving the precision of dataflow analysis originally proposed by Bruynooghe (1991). The idea is to propagate variable bindings by repeating the analysis of (parts of) the body of a clause. The results indicate that a reexecution-based analysis using a simple (and presumably less precise) abstract domain may match the performance of an analysis that uses a more elaborate abstract domain but does not use reexecution. On a simple domain of modes, reexecution led to significant improvements in precision (over 50% on the average), while incurring an additional runtime cost of only 8% on the average. On a more elaborate domain that maintained more information about term structure, the improvements in precision were not as large, while the cost in speed was a factor of about 2.3 on the average. The precision of the reexecution-based strategy using the simple domain of modes is more or less comparable to one based on a more elaborate domain that maintains structural information about terms but does not use reexecution; however, the reexecution-based analysis on the simple domain is approximately twice as fast as the analysis on the more elaborate domain that does not use reexecution.

6 Applications

6.1 Type inference

There has been a tremendous amount of research on type inference in the context of logic programming, and any kind of detailed treatment is beyond the scope of this paper: the interested reader is referred to, among others, the work of Azzoune (1988), Bruynooghe and Janssens (1988), Barbuti and Giacobazzi (1992), Fruhwirth et al. (1991), Heintze and Jaffar (1990a, 1990b, 1992a, 1992b), Horiuchi and Kanamori (1987), Klužniak (1987a), Kanamori and Horiuchi (1985), Mishra (1984), Pfenning (1992), Pyo and Reddy (1989), Reddy (1988), Xu and Warren (1988), Yardeni and Shapiro (1987), and Zobel (1987). While logic programs are usually dynamically typed, the availability of type information at compile time is useful for a variety of reasons, including data abstraction and specification of program behavior, early detection of errors, and code optimization. There appear to be two principal views of what constitutes a type: the *prescriptive* view, where types are made explicit via declarations, and the *descriptive* view, where types are implicit and obtained via program analysis. In the prescriptive view, there is no single universe of objects: instead, each type has its "domain" or "carrier", non-logical symbols come with declarations specifying their types, e.g.: "+ : **Int** × **Int** ⟶ **Int**", and rules of well-formedness of the language ensure that ill-typed constructs are syntactically disallowed. In the descriptive view, there is a universe of values and types represent certain subsets of this universe, and an object has a certain type if its value belongs to the subset of the universe corresponding to that type.

For the purposes of this discussion, we will be concerned primarily with the descriptive view of types and type inference. Here we consider only two approaches to type inference: the regular-tree based approach of Mishra (1984) and Reddy (see Pyo and Reddy 1989, Reddy 1988); and the set-based analysis of Heintze and Jaffar (1990a,1990b,1992a,1992b).

Regular tree-based type analysis: This approach is due to Mishra (1984), with later extensions to incorporate polymorphism due to Reddy (see Pyo and Reddy 1989, Reddy 1988). The work of Klužniak (1987a) and Zobel (1987) are also instances of this approach. A number of considerations arise in this context:

Notions of type : What is an appropriate notion of "type"? Stated differently, what is a natural notion of "type error" for logic programs?

Compositionality : If a predicate p is defined in terms of predicates $q_1, \ldots, q_n$, then the type of p should be defined using the *types* of $q_1, \ldots, q_n$.

Algorithmic issues : How are type equations involving recursive types (such as lists) to be solved? What are the implications with regard to efficiency?

To address the first point, Mishra (1984) proposed that the notion of type error should correspond to formulae, or goals, that always fail. In other words, a formula is ill-typed if assuming it leads to a contradiction. The second point is addressed via a set of inference rules that describe how the type of a predicate is to be determined given the types of the predicates used to define it. Types are described using *regular trees* (to simplify the manipulation of regular trees during analysis and address the third point above to some extent, most authors impose an additional requirement of *cartesian closure* that will be discussed later). Essentially, regular trees over an alphabet Σ are defined as follows: if $a \in \Sigma$ then a is a regular tree; if $S_1, \ldots, S_n$ are regular trees and f is an n-ary function symbol then $f(S_1, \ldots, S_n)$ is a regular tree; if S_1 and S_2 are regular trees then so are $S_1 + S_2$ and $S_1 \times S_2$; and if F is a "regular function" (see below) over regular trees, then $fix(F)$ is a regular tree. Here, a function is said to be regular if it can be expressed using the constructions listed, namely, using constructor symbols, sum, product, and *"fix"*. Intuitively, regular trees represent sets of terms, i.e., types: elements of the alphabet Σ denote primitive types; a regular tree $f(S_1,\ldots,S_n)$ denotes the set of terms $f(t_1,\ldots,t_n)$ where t_i is an element of the set of terms denoted by S_i, $1 \leq i \leq n$; a regular tree $S_1 + S_2$ denotes the union of the sets of terms denoted by S_1 and S_2, while $S_1 \times S_2$ denotes the cartesian product of the sets of terms denoted by S_1 and S_2; and a regular tree $fix(F)$ denotes the least fixpoint of the regular function F. Regular trees are closed under intersection, which is computable. Further, the inclusion and containment problems (and, therefore, equality) for regular trees are decidable. Regular trees can be expressed using *leaf linear grammars* (Mishra and Reddy 1985) or *regular unary Horn programs* (Yardeni and Shapiro 1987).

It turns out that keeping track of dependencies between different argument positions of predicate and function symbols during type inference leads to significant complications in the analysis algorithms and can be quite expensive computationally (see, for example, Jones and Muchnick 1981). Because of this, researchers typically ignore dependencies between terms that can appear in different argument positions of a function or predicate symbol by using *cartesian-closed* (or "tuple-distributive") regular trees. Intuitively, the idea is as follows: suppose we have a predicate defined by two clauses p(a, b) and p(c, d). Strictly speaking, there are dependencies between the different argument positions of this predicate, e.g., a appears in the first position if and only if b appears in the second position. An exact description of the type of p/2 would therefore be given by $\{\langle a,b\rangle, \langle c,d\rangle\}$. If we do not keep track of these dependencies, the type of p/2 is described as follows: "the first argument is either a or c, while the second argument is either b or d", i.e., by $\{a, c\} \times \{b, d\}$. Notice that this can introduce some imprecision, e.g., in this example, the type $\{a, c\} \times \{b, d\}$ contains the element $\langle a, d\rangle$ even though p(a, d) is not true. Thus, cartesian closure "closes off" a set by introducing new elements that allow us to ignore dependencies between argument positions.

The type inference procedure used by Mishra (1984) involves using a set of inference rules to generate a set of inclusion constraints, then solving these constraints to find the maximal solution (if one exists).

Mishra's algorithm imposes some restrictions that preclude the inference of polymorphic types. The work of Pyo and Reddy (1989) extends the work of Mishra to deal with parametric polymorphism. Pyo and Reddy consider the type of a predicate to consist of two parts: a *structural type*, which shows the structure of the arguments in all successful atoms; and an *implicational type*, which is of the form "$\forall \bar{\alpha}.p(\bar{r}) \supset p(\bar{s})$" and captures the propagation of type information between different arguments.

Set-based analysis: The set-based analysis of Heintze and Jaffar is similar in some ways to the regular-tree based approach to type inference, in that both give upper approximations to the success set of a predicate, and both require solving systems of inclusion constraints using non-iterative techniques. It is, however, a fundamentally different approach: regular tree-based type inference is based on the manipulations of objects that satisfy certain properties (e.g., cartesian closure) that affect the structure of the sets of terms that can be described. Set-based analysis, on the other hand, uses just one source of approximation: dependencies between different variables (rather than argument positions) are ignored. The difference is illustrated by the following example. Consider the following program:

```
p(a, b).
p(b, a).
```

Due to cartesian closure, a regular tree-based analysis would produce a type $\{a, b\} \times \{a, b\}$ for p/2, indicating that each argument of this predicate could have a value of either a or b. On the other hand, since there are no dependencies between any variables in this program, the set-based analysis of Heintze and Jaffar would produce the type $\{\langle a, b \rangle, \langle b, a \rangle\}$ for this predicate.

A fundamental notion in set-based analysis is that of a *set substitution*. While a substitution maps variables to terms, a set substitution maps variables to sets of terms. We can define an approximation operator *abs* that maps a set of substitutions to a set substitution, as follows: given a set of substitutions Θ, $abs(\Theta) = \alpha$ is a set substitution such that for any variable X,

$$\alpha(X) = \{t \mid \exists \theta \in \Theta : \theta(X) = t\}.$$

The approximate model-theoretic semantics of a program, as computed by Heintze and Jaffar's set-based analysis, is given by the least fixpoint $lfp(\tau_P)$ of an operator τ_P, which is defined as follows:

$$\tau_P(R) = \{a \in \alpha(H) : H \ :- \ B_1, \ldots, B_n \in P,$$
$$\alpha = abs(\{\theta : \theta(B_i) \in R\})\}.$$

It turns out that for finite programs, the approximate model-theoretic semantics so defined is finitely computable (Heintze and Jaffar 1990a,1990b), and is an upper approximation to the actual semantics of the program. The difference between the actual and approximate semantics of a program is illustrated by the following example. Consider the following program:

```
p(a, b).
p(b, a).
r(X) :- p(X, X).
q(X, Y) :- p(X, Y).
```

The exact semantics of the program is given by {p(a,b), p(b,a), q(a,b), q(b,a)} while the approximate semantics is given by {p(a,b), p(b,a), q(a,b), q(b,a), q(a,a), q(b,b)}.

Because the definition of p/2 does not involve any variables, there is no loss of precision when its approximate semantics is computed. However, when computing the approximate semantics for q/2, analysis of the clause

q(X, Y) :- p(X, Y).

produces the set substitution $\{X \mapsto \{a, b\}, Y \mapsto \{a, b\}\}$, where the dependency between the variables X and Y has been decoupled. Because of this, the approximate semantics of q/2 contains some extraneous elements that do not appear in the concrete semantics.

Notice that a regular tree-based type analysis of this program would, from the cartesian closure requirement, have produced the type {a, b} for the predicate r/1, while the set-based analysis correctly infers that the success set of r/1 is empty.

6.2 Groundness analysis

Groundness analysis is aimed at determining, at each program point, which variables are bound to ground terms. It is motivated primarily by compilation concerns: such information—also called "mode" information (Warren 1977)—can be used to generate more efficient code for unification (Van Roy et al. 1987), information about groundness of variables and (lack of) aliasing between them is used in parallelizing compilers for PROLOG (Hermenegildo et al. 1992, Jacobs and Langen 1989, Muthukumar and Hermenegildo 1991, 1992), in translators from logic programming to functional programming languages (Reddy 1984), and in reducing suspension testing in of the implementation of concurrent constraint languages (Debray 1993). It is also of historical interest in that it was the first dataflow analysis problem considered for logic programming languages (Mellish 1981).

Groundness analysis has been considered by a number of researchers: because it is conceptually so simple, it is generally given as an instance of various frameworks proposed for dataflow analyses of logic programs. Here we discuss the overall structure of groundness analysis, describe a few of the abstract domains that have been proposed for this purpose, and consider some practical issues.

Groundness analysis is in many ways the prototypical top-down analysis for logic programs. Starting with a "calling pattern" for predicates in a "top-level" query, i.e., a description of which argument positions are ground and which are free, the analysis examines the clauses of the program, propagating abstract substitutions across their bodies from left to right until all literals have been processed, at which point the resulting abstract substitution is applied to the head of the clause to obtain a "success pattern". Calling and success patterns for each predicate are recorded, and OLDT-resolution is typically used to guarantee termination. While the program can be processed in any order, it is usually more efficient to process it in depth-first order. The efficiency of the analysis can be improved by keeping just a single success pattern for each calling pattern by taking least upper bounds. For example, consider a predicate defined by the clauses

```
p(X, X, Y).
p(X, Y, X).
```

Now consider a calling pattern $\langle ground, free, free \rangle$ for this predicate. A call described by this pattern can succeed in two different ways: the first clause can succeed binding the second argument to a ground term and the third argument a free variable; and the second clause can succeed binding the second argument to a variable and the third argument to a ground term. The success patterns corresponding to these are $\langle ground, ground, free \rangle$ and $\langle ground, free, ground \rangle$. Instead of maintaining these two different success patterns for the given calling pattern, we can use a conservative summary obtained by taking their "least upper bound"—in this case, this might yield $\langle ground, any, any \rangle$.

An important concern in groundness analysis is *aliasing* between variables. If the abstract domain is substitution-closed, then—as discussed in Section 5.2, aliasing effects can safely be ignored. Examples of such abstract domains are {*ground, any*}, considered by Debray and Warren (1988) and Mannila and Ukkonen (1987). However, if the abstract domain is not substitution-closed, e.g., if it contains an element describing only unbound variables, then aliasing effects must be taken into account explicitly in order to guarantee soundness. The problem is illustrated by the following program:

```
p :- q(X, Y), r(X), s(Y).
q(U, U).
r(a).
s(_).
```

An analysis that attempts to keep track of free variables but does not take aliasing into account will not notice that because q/2 aliases its arguments together, the argument Y to s/1 is actually ground, even though Y is not explicitly bound to a ground term anywhere in the program.

For applications such as parallelization of PROLOG programs, substitution-closed abstract domains that allow aliasing to be ignored without compromising soundness are not precise enough, and it is necessary to use domains that allow aliasing and sharing between variables to be taken into account explicitly. Two domains that have recently been proposed for this, and which allow dependencies between variables to be tracked with a great deal of precision, are *Sharing*, originally proposed by Jacobs and Langen (1989) and subsequently used by Muthukumar and Hermenegildo (1991, 1992); and *Prop*, described by Marriott et al. (1990a) and discussed further by Cortesi et al. (1991). The idea behind the abstract domain *Sharing* is to describe a set of substitutions Θ using a set of sets of variables A: two variables X and Y occur in an element of A if there is some element θ in Θ such that $\theta(X)$ and $\theta(Y)$ share a variable. Thus, suppose that the set of substitutions that may be encountered at a program point is Θ, and that this is described by an abstract substitution A: then, a variable X is guaranteed to be ground at that point if X does not occur in any element of A, while two variables Y and Z are guaranteed to be independent at that point if there is no element of A that contains both Y and Z. In the approach of Marriott, Søndergaard and Jones, the abstract domain for each clause in a program is the set of propositional formulae that can be constructed from the logical constants *true* and *false* and the variables appearing in that clause, using the connectives $\wedge$ ("and"), $\vee$ ("or"), and $\Leftrightarrow$ ("if and only if"), modulo logical equivalence. A formula φ in this abstract domain describes a substitution θ if and only if the truth assignment τ, given by

$$\tau(x) = true \quad \text{iff} \quad \theta(x) \text{ is a ground term}$$

makes φ true. For example, the formula $x \Leftrightarrow y$ describes the substitutions $[x \mapsto \mathtt{a}, y \mapsto \mathtt{f(b)}]$ and $[x \mapsto z, y \mapsto g(z,z)]$, but not $[x \mapsto \mathtt{a}]$; the formula $x \wedge (y \vee z)$ describes the substitution $[x \mapsto \mathtt{f(a)}, y \mapsto u, z \mapsto \mathtt{a}]$, but not $[x \mapsto \mathtt{f(a)}, y \mapsto u, z \mapsto u]$.

Marriott and Søndergaard (1993b) discuss precise and efficient groundness analysis using such formulae. The utility of such domains is also borne out by recent experimental work by Le Charlier and Van Hentenryck, which indicates that dataflow analysis using the domain *Prop* is remarkably precise and, for most programs, not very expensive (Le Charlier and Van Hentenryck 1992b).

6.3 Occur check optimization

In theory, a variable x cannot be unified with a term such as $f(x)$ that contains that variable. Because of this, unification of a variable x with a term t should first check whether t contains x, in which case unification should fail. In practice, however, most PROLOG implementations regard this check, called the "occur check," to be too expensive, and omit it. While this seems not to pose a problem in most cases, it is possible to construct examples where this causes unsound results (a query succeeds that should have failed) or nontermination (arising from traversals of a term containing a "cycle"). Another problem is that different PROLOG systems behave very differently when confronted with terms whose unification with occur check would fail (Marriott and Søndergaard 1988b).

One way to avoid these problems without incurring the performance penalty that would result from always using unification with occurs check is to use compile-time analysis to detect situations where occur checks might be necessary, and to insert them automatically in these places during compilation. This was first suggested by Plaisted (1984), and formalized as an abstract interpretation by Søndergaard (1986). The basic intuition is that an occur check may be needed only if both the terms being unified have multiple occurrences of a variable, e.g., a goal p(X, X) being unified with a clause head p(Y, f(Y)). For analysis purposes, therefore, it is necessary to consider the occurrences of repeated variables in a goal at runtime. While the simplest case, where there are repeated occurrences of variables in the program text, is easily detectable at compile time, in general repeated variables may arise in a goal at runtime, because of substitutions, even if the program text does not contain repeated variables. This can happen in two ways: *sharing*, where two distinct variables in a goal become aliased together, e.g., in the goal p(X, Y) given the substitution $[X \mapsto Y]$; and *spawning*, where a substitution maps a variable to a term containing repeated variables, e.g., in the goal p(X) given the substitution $[X \mapsto f(Y, Y)]$.

The abstract domain of Søndergaard captures these aspects of substitutions. The abstract domain for a clause with variables V is a set of abstract substitutions, where an abstract substitution β over V is a pair $\langle G, E \rangle$ where $G \subseteq V$ represents the set of variables that are ground under every substitution in $conc(\beta)$, and $E \subseteq V \times V$ consists of two kinds of elements: elements of the form (x, y) where $x \neq y$, indicating that for some substitution θ in $conc(\beta)$, the term $\theta(x)$ may share variables with $\theta(y)$; and elements of the form (x, x), indicating that for some concrete substitution θ in $conc(\beta)$, the term $\theta(x)$ may contain multiple occurrences of a variable. Abstract unification proceeds by first propagating groundness information, then testing the terms to be unified to determine whether unification could lead to the creation of a "cyclic" term—if so, then an occurs check should be used for that unification. After this, sharing and spawning information is propagated. Testing for circularity consists of constructing a graph whose nodes are the argument positions of the two terms being unified; adding edges between argument positions that share a variable or spawn multiple occurrences of a variable, and also between corresponding argument positions of the two terms; and finally, looking for certain kinds of cycles in the resulting graph.

6.4 Compile-time garbage collection

One source of inefficiency in implementations of logic programming languages is that there is no notion of destructive update: terms must be explicitly copied when an updated version is needed.[16] This is expensive for two reasons: first, there is the cost of constructing the new copy, which can be significant if the term being copied is large; and second, there is the

---

[16]Some PROLOG systems, such as Sicstus PROLOG (Carlsson and Widen 1988), provide primitives for destructive updates, but these are not standard and do not have a logical semantics.

eventual cost of garbage collection to reclaim memory that is no longer in use.

The problem can be addressed to some extent by using compile-time analysis to identify situations where copying can be safely replaced by destructive updates without affecting the behavior of the program. This idea, which has also been considered by a number of researchers in the context of functional programming languages, was proposed in the context of logic programming languages by Vataja and Ukkonen, who described static and dynamic analyses to detect terms that are dead and whose memory can be reused (Vataja and Ukkonen 1984). Bruynooghe (1986) described a compile-time lifetime analysis algorithm. Kluźniak gave a somewhat more powerful algorithm applicable to a more restricted form of PROLOG, called "ground PROLOG", which essentially disallows partially instantiated structures (Kluźniak 1987b). The analysis was formalized as an abstract interpretation by Mulkers *et al.* (1994).

The idea underlying the approach of Mulkers *et al.* is that the storage for a term can be safely reused at a point in a program if it can be guaranteed that there are no references to that term, i.e., if the term is dead, beyond that point. Dataflow analysis is used to determine the lifetime of (components of) a structure. For this, it is necessary to take into account sharing of memory between terms. Since the standard semantics for logic programming languages does not provide information about such sharing, it is necessary to augment such semantics to express this information. The concrete representation on which the analysis of Mulkers *et al.* is based therefore consists of two components: a *term environment* that describes substitutions for variables; and a *sharing component* that describes which structures share subterms, and which of their subterms are shared.

Analogously, The abstract domain also has two components: sets of term environments are represented using a type graph, while sharing is described by an *abstract sharing* component, which is a set of pairs of type graph nodes that can possibly share memory at runtime. Abstract unification therefore must update both these components. Updating the type graph is analogous to the handling of unification in type inference, and is discussed by Bruynooghe and Janssens (1988) and Janssens (1990). The handling of the abstract sharing component requires the propagation of new sharing resulting from unification: in concrete unification this would involve taking a transitive closure of any sharing introduced by the unification, but in abstract unification an *alternating closure* (based on a proposal due to Søndergaard in his analysis of occur-check optimization (Søndergaard 1986), discussed in Section 6.3) is used in order to improve precision.

There are two components to the dataflow analysis: a *sharing analysis*, which describes, at any program point, which components of terms may share storage. For compile-time garbage collection, we also need to know which components of a structure are dead and may safely be reused. The basic idea behind lifetime analysis is to assume that the description of the top-level query specifies which components of the query are live at the end of the query. From this, it is possible to determine, for each call, which components of the call are live when the call returns. Liveness information can then be propagated as part of the analysis, and the memory for a structure can be reused after the last use of that structure, i.e., when it becomes dead.

A different approach is taken by Sundararajan *et al.*, who consider compile-time memory reuse for concurrent logic programming languages (Sundararajan *et al.* 1992). These authors restrict themselves to detecting situations where a term has a single producer and a single consumer, and which can therefore be reused by the consumer. The concrete semantics, which must account for the fact that goals are executed concurrently (and therefore that any interleaving of the primitive actions represents a legal execution), is a transition-system-based operational semantics. Unlike the analysis of Mulkers *et al.*, the abstract domain used is a very simple one, where the information maintained about each variable is simply whether or not it has a single producer and a single consumer. Experimental results on a set of small benchmarks suggest significant improvements in memory usage.

6.5 Argument size analysis

Information about relationships between the "sizes" of different variables at various points in a program is useful for a number of applications, such as termination analysis (Plümer 1990, Verschaetse and De Schreye 1991) and complexity analysis (Debray and Lin 1993). For example, a common approach to proving that any call to a (recursive) predicate always terminates is to show that the size of its arguments decreases at each recursive call. For complexity analysis, since the computational cost of a predicate typically depends on the size of its arguments, it is necessary to express the cost of a predicate as a function of its input size. The problem has been investigated by Debray and Lin (1993), Van Gelder (1990), and Verschaetse and De Schreye (1991).

The approach of Verschaetse and De Schreye expresses the "size" of a term using *semilinear norms*. A semilinear norm is a function $\|\cdot\|$ that maps terms to natural numbers, where intuitively for any term t, the value of $\|t\|$ depends only on the principal functor of t and certain "relevant" subterms of t. Since the terms appearing in a program generally contain variables whose bindings are not available until runtime, it is typically not possible to determine the exact size of a term at compile time. Instead, the analysis computes expressions that give the size of a term according to the appropriate norm, which is assumed to be supplied by the user, though Verschaetse and De Schreye (1991) and Decorte, De Schreye and Fabris (1993) show that, in many cases, the appropriate norm can be inferred from the types of argument positions.

The analysis manipulates linear size relations between argument positions of predicates. The linearity condition is imposed for efficiency reasons: of course, not all predicates have argument size relations that are linear, and in such cases it is necessary to approximate their size relations using linear relations. For analysis purposes, size relations are viewed as n-tuples of reals. Not all sets of reals are convenient, and the analysis considers only affine spaces, i.e., sets that are solutions to equations of the form

$$a_{i1}x_1 + \cdots + a_{in}x_n = c_i, \qquad 1 \leq i \leq m.$$

The analysis is based on the framework of Bruynooghe (1991). An abstract substitution is either $\bot$, or $\top$, or a triple $\langle A, \bar{X}, \bar{c} \rangle$, where A is an $m \times n$ matrix (in reduced row-echelon form), $\bar{X}$ is a column matrix with n elements $X_1, \ldots, X_n$, which forms the domain of the abstract substitution, and $\bar{c}$ is a column matrix of n reals. The various abstract operations required in Bruynooghe's framework, as discussed in Section 4.1.1, use operations on affine spaces, such as intersection, disjunction (which involves taking the union of two affine spaces and "closing up" the result to make it affine), restriction (which involves projection into a lower-dimensional space), and extension (which is an embedding into a higher-dimensional space).

6.6 Cost analysis

The ideas behind argument size analysis can be taken a step further, to estimate the computational cost of a program for a given argument size. In the context of logic programs, this was first addressed by Kaplan (1988). Because this work uses techniques from the area of term rewriting systems, it imposes syntactic restrictions on programs that preclude some interesting programs, in particular programs that construct intermediate structures. A different approach is taken by Debray and Lin (1993). This work, which gives worst case upper bound estimates to the complexity of a program, assumes that mode and data dependency analyses have been carried out, so that for each variable in the body of a clause we can determine which of its occurrences are "input" occurrences and which are "output" occurrences.

The analysis first carries out an argument size analysis to express the size of the output arguments of a predicate in terms of the size of its input arguments. This differs from the work of Verschaetse and De Schreye (1991, 1992) in that instead of using linear size relations, it constructs and solves difference equations for the size relations. Because it is not possible, in general, to solve arbitrary difference equations, the equations obtained from a program are simplified where necessary to allow solutions to be found: this can lead to a loss of precision in some cases.

After argument size analysis, each predicate is analyzed to estimate the number of solutions it may produce. This analysis uses properties of unification to deal with equality constraints of the form '$t_1 = t_2$', and graph-theoretic algorithms to deal with inequality constraints of the form $t_1 \neq t_2$, $t_1 \geq t_2$, etc. For example, given a system of disequations of the form $x \neq y$ for a set of variables that range over the same (finite) domain, a solution to the system can be characterized as a solution to a graph coloring problem, where the number of colors corresponds to the size of the underlying domain (which is assumed to have been estimated using type analysis). Estimating the number of solutions therefore amounts to estimating the chromatic polynomial of the graph. Since optimal graph coloring is an NP-complete problem, constructing the chromatic polynomial is obviously NP-hard, and the authors give a polynomial-time approximation algorithm for this problem. They also describe a different approach to estimating the number of solutions to finite-domain constraint satisfaction problems by estimating the number of cliques of a certain size in a graph derived from the problem. Recursive predicates are handled by constructing and solving difference equations.

Finally, the information about argument sizes and number of solutions is combined to determine the computational cost of a predicate in terms of the size of its input, where the measure of cost may be the number of resolutions, the number of unifications, the number of machine instructions, etc. Again, recursion is handled using difference equations, which are simplified where necessary in order to allow solutions to be computed.

7 Analysis of concurrent logic programs

The analyses considered so far have assumed that the execution strategy of the program (i.e., the literal selection strategy) is known at analysis time. This provides information about the control flow behavior of the program, which is then used to propagate dataflow information. However, things become considerably more complicated for concurrent languages, since language constructs are (implicitly) executed concurrently, and there is little control flow information available to guide an analysis. The semantics of such languages typically specify that any sequential interleaving of the primitive operations in a program constitute a legal execution. Since dataflow analyses are expected to take all possible executions of a program into account, this raises nontrivial combinatorial problems that do not occur in the sequential case. Since it is obviously not practical to consider all possible interleavings of all the primitive operations in a program, it is necessary to approximate this runtime behavior in such a way that suspension and synchronization of processes is taken into account.

The concurrent logic programming languages we consider are commonly known as *flat* languages. Examples of such languages include FCP (Houri and Shapiro 1987), FGHC (Ueda 1987), Janus (Saraswat *et al.* 1990), and Strand (Foster and Taylor 1989): Shapiro (1989) gives a more detailed discussion. Operationally, the execution of a literal causes all of the clauses for that literal to be tried concurrently. A clause can be selected for execution only if its *guard*—which consists of a set of tests—is satisfied. A guard may suspend because there is insufficient information for its tests: this is the primary synchronization mechanism in such languages. When a clause has been selected for execution, all of its body literals are executed in parallel.

The earliest work on dataflow analysis of concurrent logic programs is that of Codish et al. (1988) and Gallagher et al. (1988), who consider FCP programs. Their analysis processes clause bodies from left-to-right, ignoring synchronization and suspension behavior. Thus, while the analysis is potentially quite efficient—in particular, it sidesteps the thorny combinatorial issues arising from having to account for all possible interleavings of the primitive operations in a program—it has the limitation that certain kinds of information, e.g., about the synchronization or suspension behavior of programs, cannot be obtained. More refined analyses of concurrent logic programs have since been described by Codognet et al. (1990) and Codish et al. (1991). A scheme for dataflow analysis of concurrent logic programs for compile-time memory reuse, due to Sundararajan et al. (1992), is discussed in Section 6.4.

The approach of Codognet *et al.*: The concrete domain of Codognet et al. is based on AND-OR trees. An analysis consists of three components: an abstract domain D, containing the special components $\bot$ (denoting a complete lack of information), *fail* (denoting certain failure), and *suspend* (denoting certain suspension); an abstract operation $\downarrow\pi$ that simulates suspension checking and guard execution; and an abstract operation $\uparrow\pi$ that simulates the communication of (output) bindings to the calling literal. Two analysis algorithms are given: a "meta-algorithm" that simulates all possible computations, i.e., all possible interleavings of atomic actions; and a simpler algorithm that allows several computations to be factored into a single one under the assumption that the abstract domain satisfies a simple monotonicity property.

The "meta-algorithm" that explicitly considers every possible execution is a simple loop that, starting with the initial calling pattern for the top-level query, repeatedly processes each active goal until there are no active goals. Each active goal is processed by removing it from the active list and creating a new abstract computation tree for it, after which each leaf node in each abstract computation tree is processed. The processing of a leaf node proceeds as follows: first, the analysis simulates head matching, suspension checking, and guard execution using $\downarrow\pi$. If the guard does not fail or suspend, $\uparrow\pi$ is used to communicate output variable bindings to the caller (and brothers), after which each body literal is added as a child to the node being processed, and each such new child node is added to the set of active nodes.

There are some obvious combinatorial problems that make this algorithm impractical. Codognet *et al.* describe a simpler and more efficient algorithm that does not propagate variable bindings out to the caller as soon as possible. Instead, a "local fixpoint computation" within the clause is used to compute and propagate out together all the outputs that the body literals can produce. The idea is to process the literals in the body of a clause repeatedly from left to right: after all the body literals have been processed, the resulting abstract state is used to reanalyze the body. This is repeated until no new calling pattern can be found for any of the body literals. The analysis is sound provided that $\downarrow\pi$ satisfies the following monotonicity condition: given two abstract domain elements d_1 and d_2, for any goal G, if $d_1 \sqsubseteq d_2$ and $\downarrow\pi(G, d_2) = \textit{fail} \vee \textit{suspend}$ then $\downarrow\pi(G, d_1) = \textit{fail} \vee \textit{suspend}$. Intuitively, this condition states that if it can be inferred, given an abstract state, that a goal will definitely not be able to commit, then increasing the amount of information about the abstract state will not cause us to change our minds and infer that that goal may be able to commit. Codognet *et al.* (1990) use this approach in an anaylsis for deadlock detection for concurrent logic programs.

The approach of Codish *et al.***:** The work of Codish *et al.* on suspension analysis differs from that of Codognet *et al.* in that it uses a simpler operational semantics based on a transition system, which is abstracted directly in the analysis. The authors show that for abstract domains whose elements are "downward closed" (see below), it suffices to consider a single scheduling policy, thereby simplifying the analysis considerably. They consider a variety of abstract descriptions: that used by Codognet *et al.* for their deadlock analysis corresponds to one of these, which Codish *et al.* call "simple sharing."

If $\langle S, \leq \rangle$ is a partially ordered set and $X \subseteq S$, then X said to be *downwards closed* if and only if

$$\forall x \in X \ \forall s \in S : \text{ if } s \leq x \text{ then } s \in X.$$

An abstract interpretation with abstract domain $\mathcal{D}_{abs}$, concrete domain $\mathcal{D}_{conc}$, and concretization function *conc* is downwards closed if and only if $conc(d)$ is downwards closed for each $d \in \mathcal{D}_{abs}$. The significance of downward closed analyses, as shown by Codish *et al.* (1991), is that for such analyses it is not necessary to consider all possible scheduling policies: it suffices to consider just one. For suspension analysis, we are usually interested in possible suspension, i.e., where some execution sequence may cause suspension (rather than definite suspension, where every execution will lead to suspension). For this, it is necessary to keep track of "definite sharing" between variables, and it turns out that abstract domains for this are downwards closed.

The basic idea behind the analysis of Codish *et al.* is to construct and examine abstract transition graphs, which describe the concrete transition system semantics of programs. An abstract state is a goal, i.e., a set of literals, together with an abstract substitution describing definite sharing between variables in the goal. Nodes in the abstract transition graph are abstract states, while edges are abstract transitions, labelled with goals to indicate scheduling. If the abstract transition graph for a program does not contain a *suspend* state, then the corresponding program and initial state are guaranteed to be suspension-free.

An interesting aspect of this analysis is that because nodes in the graph represent sets of literals, it is possible to get infinite graphs even if equivalent nodes are uniquely represented, because of unbounded abstract states. As a result, termination cannot be guaranteed unless an additional layer of abstraction is used to ensure that the abstract transition graphs are always finite: Codish *et al.* use an abstraction called "star abstraction", which collapses distinct literals in an abstract state that have the same predicate symbol into a single literal. A similar problem arises also in the context of analysis of modular programs (see Section 10, and a similar solution used by Codish *et al.* (1993)).

8 Analysis of Constraint Logic Programs

Recently there has been a great deal of interest in dataflow analysis of constraint logic programs. Constraint logic programming generalizes traditional logic programming in that objects in the underlying domain of computation are not restricted to be terms in the Herbrand universe, but may be elements of an appropriate algebraic structure, and unification is generalized to constraint solving in that structure (Jaffar and Lassez, 1987). This added generality makes it possible to define constraint logic programming languages that are more powerful and expressive than traditional logic programming languages (in the sense that it is possible to directly define relationships that would require more complex encodings in a traditional logic programming language). However, the more powerful primitives available in constraint logic programming languages also complicate the task of analysing programs statically.

The earliest proposal for analysis of constraint logic programming languages is due to Marriott and Søndergaard (1990). Their idea is to instantiate the meta-language proposed by Marriott, Søndergaard and Jones (Marriott *et al.* 1990a), discussed in Section 4.1.2, with the semantics of the constraint language under consideration. Marriott and Søndergaard illustrate the approach by describing a framework where dataflow analyses of constraint languages can be defined by specifying a domain of abstract constraints and abstract operations to handle procedure calls and returns. As example analyses, they show how analyses for freeness and definiteness (analogous to groundness analyses for traditional logic programming languages) can be defined in this framework.

More recently, Bruynooghe and Janssens (1992) describe a different approach where the abstract operations need no longer be safe approximations to the corresponding concrete operations, but where an additional "propagation" step is introduced to reexecute program fragments (at the abstract level) where necessary until a safe approximation is obtained. The idea is to simplify the abstract domain and abstract operations while capturing the effects of a more sophisticated abstract domain through the propagation step: for example, an analysis using a simple abstract domain and operations that do not express dependencies between variables (aliasing being a simple special case) and reason about such dependencies might nevertheless obtain the effects of the a more powerful abstract domain that is able to capture such dependencies using propagation. Marriott and Søndergaard (1993a) discuss advantages and disadvantages of propagation-based analyses. They point out that while propagation generally improves precision for analyses based on downward-closed domains (see Section 7), it may decrease accuracy for domains that are not downward-closed; moreover, Marriott and Søndergaard point out that while in classical abstract interpretation one can prove the correctness of an entire analysis by simply proving the correctness of each of the basic abstract operations, this is no longer possible with analyses that use unsafe operations and rely on propagation to regain safety.

A more traditional approach is considered by Garcia de la Banda and Hermenegildo (1993), who discuss how the analysis framework discussed in Section 4.1.1, can be extended to deal with constraint logic programs in a very simple way, essentially by appropriately defining a notion of "abstract constraint" and defining abstract operations to capture the operations of conjunction and existential quantification of constraints. Analysis of constraint logic programming languages based on algebraic treatments of their semantics is discussed by Codognet and Filè (1992) and Giacobazzi et al. (1992). Bruynooghe and Boulanger (1993) discuss the relationships between abstract interpretation of traditional logic programming languages and constraint logic programming languages.

Several researchers have focused on specific classes of analyses for constraint logic programs. For example, Dumortier et al. (1993) investigate the inference of freeness of variables in the presence of mixed numerical and unification constraints. Garcia de la Banda et al. (1993) discuss the problem of independence of computations in constraint logic programs (while this work does not itself consider program analyses, the topic addressed is one that has been the subject of a great deal of program analysis literature in the context of traditional logic programming languages). They show that the "usual" notion of independence for logic programming languages such as Prolog cannot be lifted directly to constraint logic programming languages, and show that for the latter class of languages it is possible to define a number of different notions of independence for different kinds of applications. Marriott and Stuckey (1993) discuss a general methodology for the optimization of constraint logic programs based on three steps: refinement, where new constraints are added to clauses; removal, which involves adding instructions for the (optional) removal of a previously added constraint; and the reordering of constraint addition and removal operations when this is advantageous. They describe analyses to support each of the optimizations considered.

A number of authors have investigated specific analyses of $CLP(\mathcal{R})$ programs for program optimization purposes. For example, Jørgensen et al. (1991) discuss analyses for a variety of optimizations, including the detection of "trivial" constraints, where the full power of the constraint solver is not needed; mutual exclusion of clauses; and "future redundant" constraints, where a constraint that has been tested for satisfiability need not be added to the current constraint. Hanus (1993) discusses an analysis of $CLP(\mathcal{R})$ programs to detect situations where all delayed nonlinear constraints can be guaranteed to become linear at runtime. The abstract domain of this analysis uses dependencies that are conceptually very similar to functional dependencies in database theory, and normalization rules on dependencies that are closely related to Armstrong's axioms for functional dependencies (see, for example, Maier 1983). Macdonald et al. (1993) discuss the detection and removal of redundant variables $CLP(\mathcal{R})$ programs.

9 What information do implementors really want?

The discussion up to this point, which has focused on various frameworks for, and examples of, dataflow analyses for logic programs, is in some sense a description of the kinds of information that dataflow analyzers are prepared to provide. It is interesting to approach the question from the other side and ask what kinds of information users of dataflow analyses would like to have. Since one of the primary uses of dataflow analysis is in compile-time program optimization, we consider here the kinds of information that compiler writers for logic programming languages have found useful. However, it should be noted that program optimization is not the only possible use for the information obtained by dataflow analysis, and moreover, this discussion is necessarily limited to implementations that have been built upto this point, and cannot foresee the needs for future implementations. Nevertheless, it is instructive to consider the needs of "real" systems.

The work of Taylor (1989a) suggests that detailed type information, together with low level details, can be very useful in optimizing programs. The information collected by Taylor associates with each variable a pair $\langle T, R \rangle$, where:

- T is a mode and/or type, and may be an atomic type, such as nil, atom, integer, etc.; a structured type, e.g., $f(t_1, \ldots, t_n)$, where the t_i are types; or a mode, such as ground, free(A, Tr) (where A is

a set of aliases and Tr a Boolean indicating whether the value has been trailed); etc.
- R is a reference chain length, and may take on the values "0," "1," "0-1," or "?" (denoting "unknown").

This kind of information is not visible at the semantic level (indeed, it depends greatly on implementation-specific decisions), and therefore is not addressed by most of the work to date on formal semantics-based dataflow analysis of logic programs.

Van Roy's work on the Aquarius PROLOG compiler also illustrates the utility of low-level information (see Van Roy and Despain 1990, Van Roy 1990). This system implements a dataflow analysis that keeps track of whether a variable is ground, whether or not a variable has been initialized, whether a variable has been (recursively) dereferenced, and whether a variable is ground and recursively dereferenced. Using this information, Van Roy implements optimizations that lead to speed improvements of about 18%, and code size improvements of about 43%, on a variety of benchmarks. A similar observation is made by Mulkers et al. (1994), who consider dataflow analyses to detect which components of a structure are dead at a program point and may therefore be safely reused—they note that patterns of sharing that may be observed at a program point necessarily depend on low-level implementation details that are not accounted for in high-level semantic descriptions of PROLOG.

The work of Hermenegildo on &-PROLOG, an AND-parallel PROLOG system (Hermenegildo and Greene 1991, Hermenegildo et al. 1992) indicates that for certain applications, it may not be necessary to consider such low-level issues to get good performance improvements. Hermenegildo's compiler, which relies only on variable groundness and aliasing information to parallelize PROLOG programs, gives significant performance improvements for a wide variety of programs (Gianotti and Hermenegildo 1991, Hermenegildo and Greene 1991, Hermenegildo et al. 1992). Similarly, work on program specialization indicates that analyses based on high-level semantics suffice for such applications (Gallagher et al. 1988, Jacobs et al. 1990, Winsborough 1989). The conclusion—which is not entirely surprising—is that if we are concerned with applications where the information necessary is of a high level nature, then it is enough to have analyses based on high-level semantics. However, the work of Taylor and Van Roy indicates that such analyses may not be adequate for a variety of applications that are intimately concerned with low-level implementational details. To date, most of the work on formal aspects of dataflow analysis of logic programs has focused on relatively "clean" and abstract high-level semantics that are mathematically elegant, but which may not always address the needs of system implementors. However, it is becoming more and more clear that if the theory is to find applications in such contexts, it will be necessary to "dirty up" these clean semantic formulations and address low level aspects of implementations.

10 Inter-module analysis

All of the analyses discussed so far assume that the entire program is available for inspection during analysis. This assumption may not realistic, however, for large programs that are being developed in a modular style, because the resource requirements for analyzing the entire program may be prohibitive, or not all of the modules comprising the program may be available when we want to analyze a particular module. One solution is to make worst-case assumptions about any procedure not defined in the module being analyzed, but this *ad hoc* approach is esthetically unsatisfactory, and may also produce unacceptably imprecise results in practice.

An alternative approach is proposed by Codish *et al.* (1993b), based on the semantics for "open logic programs" proposed by Bossi *et al.* (1992). The essential idea here is to consider predicates that are imported from another module to be *open*, and define the meaning of a predicate as the result of unfolding the non-open literals in its clauses repeatedly: in the limit, this yields the meaning of a predicate in a module as a (possibly infinite) set of clauses whose bodies consist of (a possibly unbounded set of) open predicates only. Given a program P, let this process of repeated unfolding to a fixpoint be denoted by an operator $\mathcal{F}$, so that the meaning of a program P is given by $\mathcal{F}(P)$. The notion of composition of two open programs P_1 and P_2 is also given in terms of $\mathcal{F}$: the meaning of the composition of P_1 and P_2 is

$$\mathcal{F}(P_1 \cup P_2) = \mathcal{F}(\mathcal{F}(P_1) \cup \mathcal{F}(P_2)).$$

For program analysis purposes, it therefore suffices to abstract the iterated unfolding operator $\mathcal{F}$. In general, such abstraction may cause a loss of precision, so that given an "abstract unfolding" operator $\widehat{\mathcal{F}}$, we can guarantee only that

$$\widehat{\mathcal{F}}(P_1 \cup P_2) \sqsubseteq \widehat{\mathcal{F}}(\widehat{\mathcal{F}}(P_1) \sqcup \widehat{\mathcal{F}}(P_2)).$$

Codish *et al.* (1993b) describe a particular abstract domain and abstract composition operator: each element of the abstract domain is a pair, consisting of a clause (produced by unfolding) and an abstract substitution. An interesting technical problem arises in this context: because the clause bodies produced by repeated unfolding can be arbitrarily large, the abstract domain has infinite ascending chains, and termination cannot be guaranteed (a similar problem arises in the suspension analysis described by Codish *et al.* (1991). This requires an additional layer of abstraction that is orthogonal to the "usual" abstraction that produces finite abstractions of arbitrarily large sets of substitutions. A solution proposed by Codish *et al.* (1993b) is to use a notion of "star abstraction," introduced by Codish *et al.* (1991), where multiple literals in a clause body with the same predicate symbol are collapsed into a single literal. However, this may introduce some loss in precision. Situations where modules may be analyzed without such problems are also considered, namely, where the dependencies between modules are hierarchical, so that the modules may be analyzed in topological order; and where there are sufficient conditions for ensuring that abstract unfolding does not produce unbounded clause bodies.

Acknowledgements: This work was supported in part by the National Science Foundation under grant number CCR-8901283.

Bibliography

1. Aho, A. V., Hopcroft, J. E., and Ullman, J. D. (1974). *The Design and Analysis of Computer Algorithms.* Addison-Wesley, Reading, MA.
2. Aho, A. V., Sethi, R., and Ullman, J. D. (1986). *Compilers – Principles, Techniques and Tools.* Addison-Wesley, Reading, MA.
3. Azzoune, H. (1988). Type Inference in PROLOG. *Proc. International Conference on Automated Deduction.*
4. Bancilhon, F., Maier, D., Sagiv, Y., and Ullman, J. D. (1986). Magic Sets and other strange ways to implement logic programs. *Proc. ACM Symp. on Principles of Database Systems*, 1–15.
5. Bansal, A. K. and Sterling, L. (1988). An abstract interpretation scheme for logic programs based on type expression. *Proc. International Conference on Fifth Generation Computer Systems*, 422–429. Institute for New Generation Computer Technology (ICOT), Tokyo, Japan.
6. Barbuti, R. and Giacobazzi, R. (1992). A bottom-up polymorphic type inference in logic programming. *Science of Computer Programming*, 19(3): 281–313.
7. Barbuti, R., Giacobazzi, R. and Levi, G. (1993). A general framework for semantics-based bottom-up abstract interpretation of logic programs. *ACM Transactions on Programming Languages and Systems*, 15(1): 133–181.
8. Birkhoff, G. (1940). *Lattice Theory.* AMS Colloquium Publications vol. 25.
9. Bossi, A., Gabbrielli, M., Levi, G., and Meo, M. (1992). Contributions to the semantics of open logic programs. *Proc. International Conference on Fifth Generation Computer Systems*, 570–580. Institute for New Generation Computer Technology (ICOT), Tokyo, Japan.
10. Bruynooghe, M. (1986). Compile-time garbage collection, or how to transform programs in an assignment-free language into code with assignments. *Proc. IFIP TC2 Working Conference on Program Specifica-*

tion and Transformation.

11. Bruynooghe, M. (1991). A practical framework for the abstract interpretation of logic programs. *J. Logic Programming*, 10(2): 91–124.
12. Bruynooghe, M., and Boulanger, D. (1993). Abstract Interpretation for (Constraint) Logic Programming. Technical Report CW 183, Dept. of Computer Science, Katholieke Universiteit Leuven, Belgium.
13. Bruynooghe, M. and Janssens, G. (1988). An instance of abstract interpretation integrating type and mode inferencing. *Proc. Fifth International Conference on Logic Programming*, 669-683. MIT Press, Cambridge, MA.
14. Bruynooghe, M., and Janssens, G. (1992). Towards a framework for the abstract interpretation of constraint logic programs. *Proc. Workshop on Logic Programming Synthesis and Transformation*, Lecture Notes in Computer Science. Springer-Verlag, Berlin.
15. Bruynooghe, M., Janssens, G., Callebaut, A., and Demoen, B. (1987). Abstract interpretation: towards the global optimisation of PROLOG programs. *Proc. 1987 IEEE Symposium on Logic Programming*, 192–204. IEEE Computer Society Press, New York.
16. Carlsson, M. and Widen, J. (1988). *SICStus PROLOG User's Manual*. Swedish Institute of Computer Science, Kista, Sweden.
17. Codish, M. (1986). Personal communication.
18. Codish, M., and Demoen, B. (1993). Analysing logic programs using "Prop"-ositional logic programs and a magic wand. *Proc. 1993 International Symposium on Logic Programming*, 114–129. MIT Press, Cambridge, MA.
19. Codish, M., Gallagher, J., and Shapiro, E. (1988). Using safe approximations of fixed points for analysis of logic programs. *Meta-Programming in Logic Programming* (Selected papers from META-88). MIT Press, Cambridge, MA.
20. Codish, M., Falaschi, M. and Marriott, K. (1991). Suspension analysis for concurrent logic programs. *Proc. Eighth International Conference on Logic Programming*, 331–345. MIT Press, Cambridge, MA.
21. Codish, M., Bruynooghe, M., Garcia de la Banda, M., and Hermenegildo, M. (1993a). Top-down vs. Bottom-up Analysis of Logic Programs—Closing the Circle. Technical Report CW 177, Dept. of Computer Science, Katholieke Universiteit Leuven, Belgium.
22. Codish, M., Debray, S. K., and Giacobazzi, R. (1993b). Compositional analysis of modular logic programs. *Proc. ACM Symp. on Principles of Programming Languages*, 451–464. ACM Press, New York.
23. Codish, M., Dams, D., and Yardeni, E. (1994). Bottom-up abstract interpretation of logic programs. *Theoretical Computer Science*, 124(1):93–

125.

24. Codognet, P., and Filè, G. (1992). Computations, abstractions, and constraints. *Proc. Fourth International Conference on Computer Languages*. IEEE Computer Society Press, New York.

25. Codognet, C., Codognet, P., and Corsini, M. (1990). Abstract interpretation of concurrent logic languages. *Proc. 1990 North American Conference on Logic Programming*, 215–232. MIT Press, Cambridge, MA.

26. Corsini, M. and Filé, G. (1988). The Abstract Interpretation of Logic Programs: A General Algorithm and its Correctness. Research Report, Dept. of Mathematics, University of Padova.

27. Cortesi, A., Filé, G. and Winsborough, W. (1991). Prop revisited: propositional formulas as abstract domain for groundness analysis. *Proc. Sixth IEEE Symp. on Logic in Computer Science*, 322–327.

28. Cousot, P. (1981). Semantic foundations of program analysis. In Muchnick, S. S. and Jones, N.D. eds., *Program Flow Analysis: Theory and Applications*. Prentice-Hall, Englewood Cliffs, NJ.

29. Cousot, P. and Cousot, R. (1977). Abstract interpretation: a unified lattice model for static analysis of programs by construction or approximation of fixpoints. *Proc. Fourth ACM Symposium on Principles of Programming Languages*, 238-252. ACM Press, New York.

30. Cousot, P. and Cousot, R. (1979). Systematic design of program analysis frameworks. *Proc. Sixth ACM Symposium on Principles of Programming Languages*, 269-282. ACM Press, New York.

31. Cousot, P. and Cousot, R. (1991). Comparing the Galois Connection and Widening/Narrowing Approaches to Abstract Interpretation. Research Report, École Polytechnique, France.

32. Cousot, P. and Cousot, R. (1992). Abstract interpretation and application to logic programs. *J. Logic Programming*, 13(2&3): 103–179.

33. Debray, S. K. (1988). Static analysis of parallel logic programs. *Proc. Fifth International Conference on Logic Programming*, 711–732. MIT Press, Cambridge, MA.

34. Debray, S. K. (1989). Static inference of modes and data dependencies in logic programs. *ACM Transactions on Programming Languages and Systems*, 11(3): 419–450.

35. Debray, S. K. (1992a). On the complexity of dataflow analysis of logic programs. *Proc. Nineteenth International Colloquium on Automata, Languages, and Programming*. Lecture Notes in Computer Science vol. 623, 509–520. Springer-Verlag, Berlin.

36. Debray, S. K. (1992b). Efficient dataflow analysis of logic programs. *Journal of the ACM*, 39(4): 949–984.

37. Debray, S.K. (1993). QD-Janus: a sequential implementation of Janus

in PROLOG. *Software—Practice and Experience*, 23(12): 1337–1360.
38. Debray, S. K. and Lin, N. (1993). Cost analysis of logic programs. *ACM Transactions on Programming Languages and Systems*, 15(5): 826–875.
39. Debray, S. K. and Mishra, P. (1988). Denotational and operational semantics for PROLOG. *J. Logic Programming*, 5(1): 61–91.
40. Debray, S. K. and Ramakrishnan, R. (1993). Abstract interpretation of logic programs using magic transformations. *J. Logic Programming*, 18(2): 149–176.
41. Debray, S. K. and Warren, D. S. (1988). Automatic mode inference for logic programs, *J. Logic Programming*, 5(3): 207–229.
42. Decorte, S., De Schreye, D., and Fabris, M. (1993). Automatic inference of norms: a missing link in automatic termination analysis. *Proc. 1993 International Symposium on Logic Programming*, 420–436. MIT Press, Cambridge, MA.
43. Dietrich, S. W. (1987). Extension tables: memo relations in logic programming. *Proc. 1987 Symposium on Logic Programming*, 264–272. IEEE Computer Society Press, New York.
44. Dumortier, V., Janssens, G., Bruynooghe, M., and Codish, M. (1993). Freeness analysis in the presence of numerical constraints. *Proc. Tenth International Conference on Logic Programming*, 100–115. MIT Press, Cambridge, MA.
45. Englebert, V., Le Charlier, B., Roland, D., and Van Hentenryck, P. (1991). Generic abstract interpretation algorithms for PROLOG: two optimization techniques and their experimental evaluation. *Software Practice and Experience* 23(4).
46. Falaschi, M., Levi, G., Martelli, M., and Palamidessi, C. (1989). Declarative modelling of the operational semantics of logic languages. *Theoretical Computer Science*, 69(3): 289–318.
47. Fitting, M. (1985). A Kripke Kleene semantics for logic programs. *J. Logic Programming*, 2(4): 295–312.
48. Foster, I. and Taylor, S. (1989). Strand: a practical parallel programming tool. *Proc. 1989 North American Conference on Logic Programming*, 497-512. MIT Press, Cambridge, MA.
49. Fruhwirth, T., Shapiro, E., Vardi, M., and Yardeni, E. (1991). Logic programs as types for logic programs. *Proc. Sixth IEEE Conference on Logic in Computer Science*, 300–309.
50. Gallagher, J., Codish, M., and Shapiro, E. (1988). Specialisation of PROLOG and FCP programs using abstract interpretation. *New Generation Computing*, 6: 159–186.
51. Garcia de la Banda, M., and Hermenegildo, M. (1993). A practical approach to the global analysis of CLP programs. *Proc. 1993 Interna-*

tional Symposium on Logic Programming, 437–455. MIT Press, Cambridge, MA.

52. Garcia de la Banda, M., Hermenegildo, M., and Marriott, K. (1993). Independence in constraint logic programs. *Proc. 1993 International Symposium on Logic Programming*, 130–146. MIT Press, Cambridge, MA.
53. Giacobazzi, R. and Ricci, L. (1990). Pipeline optimizations in AND-parallelism by abstract interpretation. *Proc. Seventh International Conference on Logic Programming*, 291–305. MIT Press, Cambridge, MA.
54. Giacobazzi, R. and Ricci, L. (1992). Detecting determinate computations by a bottom-up abstract interpretation. *Proc. 1992 European Symposium on Programming*. Lecture Notes in Computer Science vol. 582, 167–181. Springer-Verlag, Berlin.
55. Giacobazzi, R., Debray, S. K., and Levi, G. (1992). A Generalized Semantics for Constraint Logic Programs. *Proc. 1992 International Conference on Fifth Generation Computer Systems*, 581–591. Institute for New Generation Computer Technology (ICOT), Tokyo, Japan.
56. Giannotti, F. and Hermenegildo, M. (1991). A technique for recursive invariance detection and selective program specialization. *Proc. Third International Symposium on Programming Language Implementation and Logic Programming*. Springer-Verlag, Berlin.
57. Hanus, M. (1993). Analysis of nonlinear constraints in CLP($\mathcal{R}$). *Proc. Tenth International Conference on Logic Programming*, 83–99. MIT Press, Cambridge, MA.
58. Heintze, N. and Jaffar, J. (1990a). A finite presentation theorem for approximating logic programs. *Proc. Seventeenth ACM Symposium on Principles of Programming Languages*, 197–209. ACM Press, New York.
59. Heintze, N. and Jaffar, J. (1990b). A decision procedure for a class of Herbrand set constraints. *Proc. Fifth IEEE Conference on Logic in Computer Science*, 42–51.
60. Heintze, N. and Jaffar, J. (1992a). Semantic types for logic programs. In Pfenning, F., ed., *Types in Logic Programming*. MIT Press, Cambridge, MA.
61. Heintze, N. and Jaffar, J. (1992b). An engine for logic program analysis. *Proc. Seventh IEEE Conference on Logic in Computer Science*.
62. Hermenegildo, M. and Greene, K. (1991). The &-PROLOG system: exploiting independent AND-parallelism. *New Generation Computing*, 9(3–4): 233–257.
63. Hermenegildo, M., Warren, R., and Debray, S. K. (1992). Global flow analysis as a practical compilation tool. *J. Logic Programming*, 13(4): 349–366.
64. Horiuchi, K. and Kanamori, T. (1987). Polymorphic type inference in

PROLOG by abstract interpretation. *Proc. Logic Programming Conference*, 107–116.

65. Houri, A. and Shapiro, E. (1987). A sequential abstract machine for Flat Concurrent PROLOG. In Shapiro, E., ed., *Concurrent PROLOG: Collected Papers*, 513-574. MIT Press, Cambridge, MA.

66. Jacobs, D. and Langen, A. (1989). Accurate and efficient approximation of variable aliasing in logic programs. *Proc. 1989 North American Conference on Logic Programming*, 154–165. MIT Press, Cambridge, MA.

67. Jacobs, D., Langen, A., and Winsborough, W. (1990). Multiple specialization of logic programs with runtime tests. *Proc. Seventh International Conference on Logic Programming*, 717–731. MIT Press, Cambridge, MA.

68. Jaffar, J. and Lassez., J.-L. (1987). Constraint logic programming. *Proc. Fourteenth ACM Symposium on Principles of Programming Languages*, 111–119. ACM Press, New York.

69. Janssens, G. (1990). Deriving Run-time Properties of Logic Programs by means of Abstract Interpretation. PhD Dissertation, Dept. of Computer Science, Katholieke Universiteit Leuven, Belgium.

70. Jones, N. D. and Muchnick, S. S. (1981). Complexity of flow analysis, inductive assertion synthesis, and a language due to Dijkstra. In Muchnick, S. S. and Jones, N. D., eds., *Program Flow Analysis: Theory and Applications*, 381-393. Prentice Hall, Englewood Cliffs, NJ.

71. Jones, N. D. and Mycroft, A. (1984). Stepwise development of operational and denotational semantics for PROLOG. *Proc. 1984 International Symposium on Logic Programming*. IEEE Computer Society Press, New York.

72. Jones, N. D. and Mycroft, A. (1986). Data flow analysis of applicative programs using minimal function graphs. *Proc. Thirteenth ACM Symposium on Principles of Programming Languages*, 296–306. ACM Press, New York.

73. Jones, N. D. and Søndergaard, H. (1987). A semantics-based framework for the abstract interpretation of PROLOG. In Abramsky, S., and Hankin, C., eds., *Abstract Interpretation of Declarative Languages*. Ellis Horwood, Chichester.

74. Jørgensen, N., Marriott, K. and Michaylov, S. (1991). Some global compile-time optimizations for CLP($\mathcal{R}$). *Proc. 1991 International Symposium on Logic Programming*, 420–434. MIT Press, Cambridge, MA.

75. Kanamori, T. (1993). Abstract interpretation based on Alexander templates. *J. Logic Programming* 15(1 & 2): 31–54.

76. Kanamori, T. and Horiuchi, K. (1985). Type inference in PROLOG and its application. *Proc. Ninth International Joint Conference on Artificial*

Intelligence, 704–707.
77. Kanamori, T. and Kawamura, T. (1987). Analyzing Success Patterns of Logic Programs by Abstract Hybrid Interpretation. Manuscript, Mitsubishi Electric Corp., Japan.
78. Kanamori, T. and Kawamura, T. (1993). Abstract interpretation based on OLDT resolution. *J. Logic Programming*, 15(1 & 2): 1–30.
79. Kanamori, T., Horiuchi, K., and Kawamura, T. (1987a). Detecting Functionality of Logic Programs Based on Abstract Hybrid Interpretation. ICOT Technical Report TR-331.
80. Kanamori, T., Kawamura, T., and Horiuchi, K. (1987b). Detecting Termination of Logic Programs Based on Abstract Hybrid Interpretation. ICOT Technical Report TR-398.
81. Kaplan, S. (1988). Algorithmic complexity of logic programs. *Proc. Fifth International Conference on Logic Programming*, 780–793. MIT Press, Cambridge, MA.
82. Kemp, R. S. and Ringwood, G. A. (1990). An algebraic framework for abstract interpretation of definite programs. *Proc. 1990 North American Conference on Logic Programming*, 516–530. MIT Press, Cambridge, MA.
83. Kifer, M., Ramakrishnan, R., and Silberschatz, A. (1988). An axiomatic approach to deciding finiteness of queries in deductive databases. *Proc. 7th ACM Symp. on Principles of Database Systems*.
84. Klužniak, F. (1987a). Type synthesis for ground PROLOG. *Proc. Fourth International Conference on Logic Programming*, 788–816. MIT Press, Cambridge, MA.
85. Klužniak, F. (1987b). Compile-time Garbage Collection for Ground PROLOG. Draft Report, Institute of Informatics, Warsaw University.
86. Langen, A. (1991). *Advanced Techniques for Approximating Variable Aliasing in Logic Programs*. PhD Dissertation, Dept. of Computer Science, University of Southern California, Los Angeles.
87. Le Charlier, B., Musumbu, K., and Van Hentenryck, P. (1991). A generic abstract interpretation algorithm and its complexity analysis. *Proc. Eighth International Conference on Logic Programming*, 64–78. MIT Press, Cambridge, MA.
88. Le Charlier, B. and Van Hentenryck, P. (1992a). Reexecution in abstract interpretation of PROLOG. *Proc. Joint International Conference and Symposium on Logic Programming*, 750–764. MIT Press, Cambridge, MA.
89. Le Charlier, B. and Van Hentenryck, P. (1992b). Groundness Analysis for PROLOG: Implementation and Evaluation of the Domain Prop. Technical Report CS-92-49, Dept. of Computer Science, Brown Univer-

sity.

90. Le Charlier, B. and Van Hentenryck, P. (1994). Experimental evaluation of a generic abstract interpretation algorithm for PROLOG. *ACM Transactions on Programming Languages and Systems*, 16(1): 35–101.

91. Macdonald, A., Stuckey, P. and Yap, R. (1993). Redundancy of variables in CLP($\mathcal{R}$). *Proc. 1993 International Symposium on Logic Programming*, 75–93. MIT Press, Cambridge, MA.

92. Maier, D. (1983). *The Theory of Relational Databases*. Computer Science Press, New York.

93. Mannila, H. and Ukkonen, E. (1987). Flow analysis of PROLOG programs. *Proc. IEEE Symposium on Logic Programming*. IEEE Computer Society Press, New York.

94. Marriott, K. and Søndergaard, H. (1988a). Bottom-up abstract interpretation of logic programs. *Proc. Fifth International Conference on Logic Programming*, 733–748. MIT Press, Cambridge, MA.

95. Marriott, K. and Søndergaard, H. (1988b). On PROLOG and the Occur Check Problem. Technical Report 88/21, Dept. of Computer Science, University of Melbourne, Parkville, Australia.

96. Marriott, K. and Søndergaard, H. (1989). Semantics-based dataflow analysis of logic programs. *Information Processing 89*, ed. Ritter, G. 601–606. North Holland, Amsterdam.

97. Marriott, K. and Søndergaard, H. (1990). Analysis of constraint logic programs. *Proc. 1990 North American Conference on Logic Programming*. 531–547. MIT Press, Cambridge, MA.

98. Marriott, K. and Søndergaard, H. (1992). Bottom-up dataflow analysis of normal logic programs. *J. Logic Programming*, 13(2& 3): 181–204.

99. Marriott, K., and Søndergaard, H. (1993a). On propagation-based analysis of logic programs. *Proc. ILPS-93 Workshop on Global Compilation*, ed. S. Michaylov and W. Winsborough, 47–65.

100. Marriott, K., and Søndergaard, H. (1993b). Precise and efficient groundness analysis for logic programs. *ACM Letters on Programming Languages and Systems* 2(1–4): 181–196.

101. Marriott, K., Søndergaard, H. and Jones, N. D. (1990a). Denotational Abstract Interpretation of Logic Programs. Manuscript, Dept. of Computer Science, University of Melbourne.

102. Marriott, K., Søndergaard, H., and P. Dart (1990b). A characterization of non-floundering logic programs. *Proc. 1990 North American Conference on Logic Programming*, 661–680. MIT Press, Cambridge, MA.

103. Marriott, K., and Stuckey, P. (1993). The 3 R's of optimizing constraint logic programs: refinement, removal and reordering. *Proc. Twentieth ACM Symposium on Principles of Programming Languages*, 334–

344. ACM Press, New York.

104. Mellish, C. S. (1981). The Automatic Generation of Mode Declarations for PROLOG Programs. DAI Research Paper 163, Dept. of Artificial Intelligence, University of Edinburgh.

105. Mellish, C. S. (1986). Abstract interpretation of PROLOG programs. *Proc. Third International Conference on Logic Programming*, Lecture Notes in Computer Science vol. 225, Springer-Verlag, Berlin.

106. Mellish, C. S. (1990). Using Specialisation to Reconstruct Two Mode Inference Systems. Manuscript, Department of Artificial Intelligence, University of Edinburgh.

107. Mishra, P. (1984). A theory of types for PROLOG. *Proc. 1984 International Symposium on Logic Programming*. IEEE Computer Society Press, New York.

108. Mishra, P. and Reddy, U. (1985). Declaration-free type checking. *Proc. ACM Symposium on Principles of Programming Languages*, 7–21. ACM Press, New York.

109. Mulkers, A., Winsborough, W., and Bruynooghe, M. (1994). Live structure dataflow analysis for PROLOG. *ACM Transactions on Programming Languages and Systems* 16(2): 205–258.

110. Muthukumar, K. and Hermenegildo, M. (1991). Combined determination of sharing and freeness of program variables through abstract interpretation. *Proc. Eighth International Conference on Logic Programming*, 49–63. MIT Press, Cambridge, MA.

111. Muthukumar, K. and Hermenegildo, M. (1992). Determination of variable dependence information through abstract interpretation. *J. Logic Programming* 13(2& 3): 315–347.

112. Nilsson, U. (1989). A Systematic Approach to Abstract Interpretation of Logic Programs. PhD Dissertation, Dept. of Computer and Information Science, Linköping University, Sweden.

113. Nilsson, U. (1991). Abstract interpretation: a kind of magic. *Proc. International Workshop on Programming Language Implementation and Logic Programming*, 299–309.

114. Pfenning, F. (1992). *Types in Logic Programming*. MIT Press, Cambridge, MA.

115. Plaisted, D. A. (1984). The occur-check problem in PROLOG. *Proc. 1984 International Symposium on Logic Programming*, 272–280. IEEE Computer Society Press, New York.

116. Plümer, L. (1990). *Termination Proofs for Logic Programs*. Lecture Notes in Artificial Intelligence, vol. 446. Springer-Verlag, Berlin.

117. Pyo, C. and Reddy, U. S. (1989). Inference of polymorphic types for logic programs. *Proc. 1989 North American Symposium on Logic Pro-

gramming, 1115–1132. MIT Press, Cambridge, MA.

118. Ramakrishnan, R. and Sudarshan, S. (1991). Top-down vs. bottom-up revisited. *Proc. 1991 International Symposium on Logic Programming*, 321-336. MIT Press, Cambridge, MA.

119. Reddy, U. S. (1984). Transformation of logic programs into functional programs. *Proc. 1984 Int. Symposium on Logic Programming*, 187–196. IEEE Computer Society Press, New York.

120. Reddy, U. S. (1988). Notions of polymorphism for predicate logic programs. *Proc. Fifth International Conference on Logic Programming.* MIT Press, Cambridge, MA.

121. Saraswat, V., Kahn, K., and Levy, J. (1990). Janus: a step towards distributed constraint programming. *Proc. 1990 North American Conference on Logic Programming*, 431-446. MIT Press, Cambridge, MA.

122. Shapiro, E. (1989). The family of concurrent logic programming languages. *Computing Surveys*, 21(3): 412–510.

123. Søndergaard, H. (1986). An application of abstract interpretation of logic programs: occur check reduction. *Proc. European Symposium on Programming*, 327–338. Lecture Notes in Computer Science vol. 213, Springer Verlag, Berlin.

124. Sundarajan, R., Sastry, A. V. S. and Tick, E. (1992). Variable threadedness analysis for concurrent logic programs. *Proc. Joint International Conference and Symposium on Logic Programming*, 493–508. MIT Press, Cambridge, MA.

125. Tamaki, H. and Sato, T. (1986). OLDT resolution with tabulation. *Proc. Third International Conference on Logic Programming*, 84–98. Springer Verlag, Berlin.

126. Tarski, A. (1955). A lattice-theoretical fixpoint theorem and its applications. *Pacific J. Math.* 5: 285–309.

127. Taylor, A. (1989a). Removal of dereferencing and trailing in PROLOG compilation. *Proc. Sixth International Conference on Logic Programming*, 48–60. MIT Press, Cambridge, MA.

128. Taylor, S. (1989b). *Parallel Logic Programming Techniques.* Prentice-Hall, Englewood Cliffs, NJ.

129. Ueda, K. (1987). Guarded Horn Clauses. In *Concurrent PROLOG: Collected Papers*, vol. 1, ed. Shapiro, E., 140-156. MIT Press, Cambridge, MA.

130. Ullman, J. D. (1989). Bottom-up beats top-down for Datalog. *Proc. ACM Symposium on Principles of Database Systems*, 140–150.

131. van Emden, M. H. and Kowalski, R. A. (1976). The semantics of predicate logic as a programming language. *Journal of the ACM*, 23(4): 733–742.

132. Van Gelder, A. (1990). Deriving constraints among argument sizes in logic programs. *Proc. Ninth ACM Symposium on Principles of Database Systems*, 47–60.
133. Van Roy, P. (1990). *Can Logic Programming Execute as Fast as Imperative Programming?* PhD Dissertation, University of California, Berkeley. (Also available as Technical Report UCB/CSD 90-600, Computer Science Divison (EECS), University of California, Berkeley.)
134. Van Roy, P. and Despain, A. M. (1990). The benefits of global dataflow analysis for an optimizing PROLOG compiler. *Proc. 1990 North American Conference on Logic Programming*, 501–515. MIT Press, Cambridge, MA.
135. Van Roy, P., Demoen, B., and Willems, Y. D. (1987). Improving the execution speed of compiled PROLOG with modes, clause selection and determinism. *Proc. TAPSOFT 1987*. Springer Verlag, Berlin.
136. Vataja, P. and Ukkonen, E. (1984). Finding temporary terms in PROLOG programs. *Proc. International Conference on Fifth Generation Computer Systems*, 275–282.
137. Verschaetse, K. and De Schreye, D. (1991). Deriving termination proofs for logic programs using abstract procedures. *Proc. Eighth International Conference on Logic Programming*, 301–315. MIT Press, Cambridge, MA.
138. Verschaetse, K. and De Schreye, D. (1992). Derivation of linear size relations by abstract interpretation. *Proc. International Conference on Programming Language Implementation and Logic Programming*.
139. Warren, D. H. D. (1977). Implementing PROLOG – Compiling Predicate Logic Programs. Research Reports 39 and 40, Dept. of Artificial Intelligence, University of Edinburgh.
140. Winsborough, W. (1987). A Minimal Function Graph Semantics for Logic Programs. Computer Science Technical Report no. 711, University of Wisconsin, Madison.
141. Winsborough, W. (1988). Automatic, Transparent Parallelization of Logic Programs at Compile Time. PhD Dissertation, Dept. of Computer Science, University of Wisconsin, Madison.
142. Winsborough, W. (1989). Path-dependent reachability analysis for multiple specialization. *Proc. North American Conference on Logic Programming*, 133–153. MIT Press, Cambridge, MA.
143. Xu, J. and Warren, D. S. (1988). A type inference system for PROLOG. *Proc. Fifth International Conference on Logic Programming*, 604–619. MIT Press, Cambridge, MA.
144. Yardeni, E. and Shapiro, E. (1987). A type system for logic programs. In *Concurrent PROLOG* vol. 2, ed. Shapiro, E., MIT Press, Cambridge, MA.

145. Zobel, J. (1987). Derivation of polymorphic types for PROLOG programs. *Proc. Fourth International Conference on Logic Programming*, 817–838. MIT Press, Cambridge, MA.

4

Modular Termination Proofs for Logic and Pure PROLOG Programs

Krzysztof R. Apt

CWI and University of Amsterdam

Dino Pedreschi

Università di Pisa

Abstract

We provide a uniform and simplified presentation of the methods of Bezem (1993) (first published as (Bezem 1989)) and of Apt and Pedreschi (1993) (first published as (Apt and Pedreschi 1990)) for proving termination of logic and PROLOG programs. Then we show how these methods can be refined so that they can be used in a modular way.

1 Introduction

1.1 Motivation

The theory of logic programming ensures us that SLD-resolution is a sound and complete procedure for executing logic programs. As a consequence, given a program P, every SLD-tree for a goal G is a complete search space for finding an SLD-refutation of G. In the actual implementations of logic programming, the critical choice is that of a tree-searching algorithm. Two basic tree-search strategies are: the *breadth-first* search which explores the tree by levels, and the *depth-first* search which explores the tree by branches. The former is a complete strategy, in the sense that it finds a success node if one exists, whereas the latter is incomplete, since success nodes can be missed if an infinite branch is explored first.

However, for efficiency reasons most implementations of logic programming adopt the depth-first strategy; in the case of PROLOG additionally a fixed selection rule is adopted. This "destroys" the completeness results linking the declarative and operational semantics of logic programming and makes it difficult to use the basic theory of logic programming for reasoning about programs.

These complications motivate research on methods for proving termination of logic programs, and in particular the approach of Bezem (1993), who proposed a method for proving termination w.r.t. all selection rules, and the approach of Apt and Pedreschi (1993), who refined Bezem's method to the leftmost selection rule of PROLOG. (For a discussion of related work the reader is referred to these two papers.)

The aim of the present paper is twofold. First, we provide a uniform and simplified presentation of these two methods, which shows that the method of Apt and Pedreschi for dealing with pure PROLOG programs is a natural extension of Bezem's method for dealing with logic programs. Secondly, we provide an extension of both methods, which supports a compositional methodology for combining termination proofs of separate programs to obtain proofs of larger programs. A number of applications are presented to substantiate the effectiveness of these modular methods in breaking down termination proofs into smaller and simpler ones, and their ability to deal with *program schemes*. In particular, simple termination proofs are exhibited for a *divide and conquer* scheme, a *generate and test* scheme, and two schemes borrowed from functional programming: a *map* scheme and a *fold* scheme.

The paper is organized as follows. In Section 2 we present the method due to Bezem (1993) for proving termination of logic programs, and in Section 3 its modification due to Apt and Pedreschi (1993) for proving termination of pure PROLOG programs. Then in Sections 4 and 5 we refine these methods so that they can be used in a modular way.

1.2 Preliminaries

Throughout this paper we use the standard notation of Lloyd (1987) and Apt (1990). In particular, for a logic program P (or simply a *program*) we denote the Herbrand Base of P by B_P and the least Herbrand model of P by M_P. Also, we use PROLOG's convention identifying, in the context of a program, each string starting with a capital letter with a variable, reserving other strings for the names of constants, terms or relations. So, for example Xs stands for a variable whereas xs stands for a term.

In the programs we use the usual list notation. The constant [ ] denotes the empty list and [. | .] is a binary function which, given a term x and a list xs, produces a new list $[x \mid xs]$ with head x and tail xs. By convention, identifiers ending with "s", like xs, will range over lists. The standard notation $[x_1, \ldots, x_n]$, for $n \geq 0$, is used as an abbreviation of $[x_1 \mid [\ldots [x_n \mid [\,]] \ldots]]$.

Throughout the paper we consider SLD-resolution and LD-resolution. The latter is obtained from SLD-resolution by using PROLOG's first-left selection rule. The concepts of LD-derivation, LD-refutation, LD-tree, etc. are defined in the usual way. By "pure PROLOG" we mean in this paper LD-resolution combined with the depth-first search in LD-trees.

By choosing variables of the input clauses and the used mgu's in a fixed way we can assume that for every program P and goal G there exists exactly one LD-tree for $P \cup \{G\}$.

In what follows we shall use the multiset ordering. A *multiset*, sometimes called *bag*, is an unordered sequence. We denote a multiset consisting of elements $a_1, \ldots, a_n$ by $bag\,(a_1, \ldots, a_n)$. Given a (non-reflexive) ordering $<$ on a set W, the *multiset ordering over* $(W, <)$ is an ordering on finite multisets of the set W. It is defined as the transitive closure of the relation in which X is smaller than Y if X can be obtained from Y by replacing an element a of Y by a finite (possibly zero) number of elements each of which is smaller than a in the ordering $<$.

In symbols, first we define the relation $\prec$ by

$$X \prec Y \text{ iff } X = Y - \{a\} \cup Z \text{ for some } a \in Y \text{ and } Z \text{ such that } b < a \text{ for } b \in Z,$$

where X, Y, Z are finite multisets of elements of W, and then define the multiset ordering $\prec_m$ over $(W, <)$ as the transitive closure of the relation $\prec$.

It is well-known (see e.g. Dershowitz (1987)) that multiset ordering over a well-founded ordering is again well-founded. In particular, the multiset ordering over the set of natural numbers with their usual ordering is well-founded.

2 Termination

2.1 Motivation

Consider the following simple program LIST:

```
list(Xs)  ←  Xs is a list.
list([H | Ts])  ←  list(Ts).
list([]).
```

It is easy to see that

- for a list t the goal ← list(t) successfully terminates,
- for a *ground* term t which is not a list, the goal ← list(t) finitely fails.

Note that in the second statement we required that t is ground. Can we drop this restriction? The answer is "No." Indeed, consider the goal ← list(X) with a variable X.

As X unifies with [H | Ts], we see that using the first clause ← list(Ts) is a resolvent of ← list(X). By repeating this procedure we obtain an infinite LD-derivation which starts with ← list(X). There is an easy fix to this problem: it suffices to reorder the clauses of the program. Then the goal ← list(X) terminates with the c.a.s. {X/[ ]}. So termination depends on the clause ordering.

Another, more interesting, possibility is to make the notion of termination independent of the clause ordering. According to this definition, a goal terminates if *all* derivations starting with it are finite. Then the goal

← list(X) does not terminate in this sense. Once the clause ordering becomes irrelevant, it is possible to adopt the view of logic programming theory and consider the program as the set (and not sequence) of clauses.

It is useful to note a simple consequence of this notion of termination. When a goal terminates in this strong sense, the corresponding computation tree is finite. Thus the depth-first search in this tree terminates, and consequently it is possible to compute by means of PROLOG *all* c.a.s.'s of the goal under consideration.

In what follows we shall study this stronger notion of termination. Our aim will be to identify for a given pure PROLOG program those goals which terminate in the above sense. Clearly, our discussion concerning the program LIST and the goal ← list(X) is equally applicable to other programs.

2.2 Terminating programs

We begin our study of termination by analyzing termination in a very strong sense, namely w.r.t. all selection rules. This notion of termination is more applicable to logic programs than to PROLOG programs. However, it is easier to handle and it will provide us with a useful basis from which a transition to the case of pure PROLOG programs will be quite natural.

In this section we study the terminating programs in the following sense.

Definition 2.1 *A program is called terminating if all its SLD-derivations starting with a ground goal are finite.*

Hence, terminating programs have the property that the SLD-trees of ground goals are finite, and any search procedure in such trees will always terminate, independently from the adopted selection rule. When studying PROLOG programs, one is actually interested in proving termination of a given program not only for all ground goals but also for a class of non-ground goals constituting the intended queries. The method of proving termination considered here will allow us to identify for each program such a class of non-ground goals. As we shall see below, many PROLOG programs, including SUM, LIST and APPEND are terminating.

To prove that a program is terminating the following concepts due to Bezem (1993) and Cavedon (1989) will play a crucial role.

Definition 2.2

- *A level mapping for a program P is a function $|\ |: B_P \to N$ of ground atoms to natural numbers. For $A \in B_P$, $|A|$ is the level of A.*
- *A clause of P is called recurrent w.r.t. a level mapping $|\ |$, if for every ground instance $A \leftarrow \mathbf{A}, B, \mathbf{B}$ of it*

$$|A| > |B|.$$

- *A program P is called recurrent w.r.t. a level mapping $|\ |$, if all its* clauses are. P is called recurrent if it is recurrent w.r.t. some level mapping.

First, following Bezem (1993), let us "lift" the concept of level mapping to non-ground atoms.

Definition 2.3
- An atom A is called bounded w.r.t. a level mapping $|\ |$, if $|\ |$ is bounded on the set $[A]$ of ground instances of A. For A bounded w.r.t. $|\ |$, we define $|A|$, the level of A w.r.t. $|\ |$, as the maximum $|\ |$ takes on $[A]$.
- A goal is called bounded w.r.t. a level mapping $|\ |$, if all its atoms are. For $G = \leftarrow A_1, \ldots, A_n$ bounded w.r.t. $|\ |$, we define $|G|$, the level of G w.r.t. $|\ |$, as the multiset bag $(|A_1|, \ldots, |A_n|)$. If $|A_i| \leq k$ for $i \in [1, n]$, we say that G is bounded by k.

The concept of boundedness is crucial when considering termination, as the following lemma shows. Recall that $\prec_m$ stands for the multiset ordering defined in the preliminaries.

Lemma 2.4 Let P be a program that is recurrent w.r.t. a level mapping $|\ |$. Let G_1 be a goal that is bounded w.r.t. $|\ |$ and let G_2 be an SLD-resolvent of G_1 from P. Then
- G_2 is bounded w.r.t. $|\ |$,
- $|G_2| \prec_m |G_1|$.

Proof. An SLD-resolvent of a goal and a clause is obtained by means of the following three operations:
- instantiation of the goal,
- instantiation of the clause,
- replacement of an atom, say H, of a goal by the body of a clause whose head is H.

Thus the lemma is an immediate consequence of the fact that an instance of a recurrent clause w.r.t. $|\ |$ is recurrent w.r.t. $|\ |$, and the following claims in which we refer to the given level mapping. ∎

Claim 1 *An instance G' of a bounded goal G is bounded and $|G'| \preceq_m |G|$.*

Proof. It suffices to note that an instance A' of a bounded atom A is bounded and $|A'| \leq |A|$. ∎

Claim 2 *For every recurrent clause $H \leftarrow \mathbf{B}$, if $\leftarrow H$ is bounded, then $\leftarrow \mathbf{B}$ is bounded and $|\leftarrow \mathbf{B}| \prec_m |\leftarrow H|$.*

Proof. Consider an atom C occurring in a ground instance of $\leftarrow \mathbf{B}$. Then it occurs in the body of a ground instance of $H \leftarrow \mathbf{B}$, say $H\theta \leftarrow \mathbf{B}\theta$. By the recurrence of $H \leftarrow \mathbf{B}$ we get $|C| < |H\theta|$, so $|C| < |H|$. This proves the claim. ∎

Claim 3 *For every recurrent clause $H \leftarrow \mathbf{B}$ and sequences of atoms $\mathbf{A}$ and $\mathbf{C}$, if $\leftarrow \mathbf{A}, H, \mathbf{C}$ is bounded, then $\leftarrow \mathbf{A}, \mathbf{B}, \mathbf{C}$ is bounded and $|\leftarrow \mathbf{A}, \mathbf{B}, \mathbf{C}| \prec_m |\leftarrow \mathbf{A}, H, \mathbf{C}|$.*

Proof. Immediate by Claim 2 and the definition of the multiset ordering. ∎

The following conclusions are now immediate.

Corollary 2.5 *Let P be a recurrent program and G a bounded goal. Then all SLD-derivations of $P \cup \{G\}$ are finite.*

Proof. The multiset ordering is well-founded. ∎

Corollary 2.6 *Every recurrent program is terminating.*

Proof. Every ground goal is bounded. ∎

These corollaries can be easily applied to various PROLOG programs. The level mapping can be usually defined as a simple function of the terms of the ground atom. The following natural concept, due to Ullman and Van Gelder (1988), will often be useful.

Define by induction a function $|\ |$, called *listsize*, which assigns natural numbers to ground terms:

$$|[x|xs]| = |xs| + 1,$$
$$|f(x_1, \ldots, x_n)| = 0 \text{ if } f \neq [\,.\,|\,.\,].$$

Note that for a list xs, $|xs|$ equals its length.

For goals with one atom it is often easy to establish boundedness by proving a stronger property.

Definition 2.7 *Let $|\ |$ be a level mapping. An atom A is called rigid w.r.t. $|\ |$ if $|\ |$ is constant on the set $[A]$ of ground instances of A.*

Obviously, rigid atoms are bounded.

Example 1

(i) Consider the program LIST. Define

$$|\mathtt{list(t)}| = |t|.$$

It is straightforward to see that LIST is recurrent w.r.t. $|\ |$ and that for a list t, the atom list(t) is rigid w.r.t. $|\ |$. By Corollary 2.6 we conclude that LIST is terminating and by Corollary 2.5 we conclude that for a list t, all SLD-derivations of LIST $\cup$ { $\leftarrow$ list(t)} are finite.

(ii) Consider now the program MEMBER:

member(Element, List) ← Element is an element of the list List.

member(X, [Y | Xs]) ← member(X, Xs).
member(X, [X | Xs]).

Using the level mapping

$$|\text{member}(x, y)| = |y|$$

we conclude by Corollary 2.6 that MEMBER is terminating and by Corollary 2.5 that for a list t, all SLD-derivations of MEMBER $\cup$ { $\leftarrow$ member(s, t)} are finite.

We now prove the converse of Corollary 2.6. With a goal G we associate the set of SLD-derivations of $P \cup \{G\}$. These SLD-derivations can be structured as a tree which we call an *S-tree* for $P \cup \{G\}$. In this tree the resolvents of a goal w.r.t. all selection rules and all input clauses constitute its direct descendants.

Lemma 2.8 *An S-tree for $P \cup \{G\}$ is finite iff all SLD-derivations of $P \cup \{G\}$ are finite.*

Proof. By the fact that we fixed the choice of mgu's and the fact that logic programs are finite, the S-trees are finitely branching. The claim now follows by König's lemma (König 1927). ∎

This lemma allows us to concentrate on S-trees. For a program P and a goal G, we denote by $nodes_P(G)$ the number of nodes in the S-tree for $P \cup \{G\}$.

Lemma 2.9 *Let P be a program and G a goal such that the S-tree for $P \cup \{G\}$ is finite. Then*

(i) for all substitutions θ, $nodes_P(G\theta) \leq nodes_P(G)$,

(ii) for all atoms A of G, $nodes_P(\leftarrow A) \leq nodes_P(G)$,

(iii) for all non-root nodes H in the S-tree for $P \cup \{G\}$, $nodes_P(H) < nodes_P(G)$.

Proof
(i) By the lifting lemma (see (Lloyd 1987)) we conclude that to every SLD-derivation of $P \cup \{G\theta\}$ with input clauses $C_1, C_2, \ldots$, there corresponds an SLD-derivation of $P \cup \{G\}$ with variants of input clauses $C_1, C_2, \ldots$, of the same or larger length. This implies the claim.
(ii), (iii) Immediate by the definition. ∎

We can now prove the desired result.

Theorem 2.10 *Let P be a terminating program. Then for some level mapping $|\ |$*

(i) P is recurrent w.r.t. $|\ |$,

(ii) for every goal G, G is bounded w.r.t $|\ |$ iff all SLD-derivations of $P \cup \{G\}$ are finite.

Proof. Define the level mapping by putting for $A \in B_P$

$$|A| = nodes_P(\leftarrow A).$$

Since P is terminating, by Lemma 2.8 this level mapping is well defined. First we prove one implication of (ii).

(ii1) Consider a goal G such that all SLD-derivations of $P \cup \{G\}$ are finite. We prove that G is bounded by $nodes_P(G)$ w.r.t. $|\ |$.

To this end take a ground instance $\leftarrow A_1, \ldots, A_n$ of G and $i \in [1, n]$. We have

$$\begin{aligned}
& nodes_P(G) \\
\geq\ & \{\text{Lemma 2.9 (i)}\} \\
& nodes_P(\leftarrow A_1, \ldots, A_n) \\
\geq\ & \{\text{Lemma 2.9 (ii)}\} \\
& nodes_P(\leftarrow A_i) \\
=\ & \{\text{definition of } |\ |\} \\
& |A_i|,
\end{aligned}$$

which proves the claim.

(i) We prove that P is recurrent w.r.t. $|\ |$. Take a clause $A \leftarrow B_1, \ldots, B_n$ in P and its ground instance $A\theta \leftarrow B_1\theta, \ldots, B_n\theta$. We need to show that

$$|A\theta| > |B_i\theta| \text{ for } i \in [1, n].$$

We have $A\theta\theta \equiv A\theta$, so $A\theta$ and A unify. Let $\mu = \mathrm{mgu}(A\theta, A)$. Then $\theta = \mu\delta$ for some δ. By the definition of SLD-resolution, $\leftarrow B_1\mu, \ldots, B_n\mu$ is an SLD-resolvent of $\leftarrow A\theta$.

Then for $i \in [1, n]$

$$\begin{aligned}
& |A\theta| \\
=\ & \{\text{definition of } |\ |\} \\
& nodes_P(\leftarrow A\theta) \\
>\ & \{\text{Lemma 2.9 (iii), } \leftarrow B_1\mu, \ldots, B_n\mu \text{ is a resolvent of } \leftarrow A\theta\} \\
& nodes_P(\leftarrow B_1\mu, \ldots, B_n\mu) \\
\geq\ & \{\text{part (ii1), with } G := \leftarrow B_1\mu, \ldots, B_n\mu \text{ and } A_i := B_i\theta\} \\
& |B_i\theta|.
\end{aligned}$$

(ii2) Consider a goal G which is bounded w.r.t. $|\ |$. Then by (i) and Corollary 2.5 all SLD-derivations of $P \cup \{G\}$ are finite. ∎

Corollary 2.11 *A program is terminating iff it is recurrent.*

Proof. By Corollary 2.6 and Theorem 2.10. ∎

2.3 Examples

2.3.1 Subset

Consider the following program SUBSET:

subset(Xs, Ys) ←
    each element of the list Xs is a member of the list Ys.

subset([X | Xs], Ys) ← member(X, Ys), subset(Xs, Ys).
subset([], Ys.

augmented by the MEMBER program.

To prove that SUBSET is recurrent we use the following level mapping:

$$|\text{member}(x, xs)| = |xs|,$$
$$|\text{subset}(xs, ys)| = |xs| + |ys|.$$

By Corollary 2.6 SUBSET is terminating and consequently by Corollary 2.5 if xs and ys are lists, all SLD-derivations of SUBSET ∪ { ← subset(xs, ys)} are finite.

In general, various choices for the level mapping exist and for each choice different conclusions can be drawn. The following three simple examples illustrate this point.

2.3.2 Append

Consider the program APPEND:

app(Xs, Ys, Zs) ←
    Zs is the result of concatenating the lists Xs and Ys.

app([X | Xs], Ys, [X | Zs]) ← app(Xs, Ys, Zs).
app([], Ys, Ys).

It is easy to check that APPEND is recurrent w.r.t. the level mapping |app(xs, ys, zs)| = |xs| and also w.r.t. the level mapping |app(xs, ys, zs)| = |zs|. In each case we get different class of goals which are bounded. The level mapping

$$|\text{app}(xs, ys, zs)| = \min(|xs|, |zs|)$$

combines the advantages of both of them. APPEND is easily seen to be recurrent w.r.t. this level mapping and if xs is a list or zs is a list, app(xs, ys, zs) is bounded (though not rigid). By Corollary 2.6 APPEND is terminating and by Corollary 2.5 if xs is a list or zs is a list, all SLD-derivations of APPEND ∪ { ← app(xs, ys, zs)} are finite.

2.3.3 Select

Consider the program SELECT:

select(X, Xs, Zs) ←

Zs is the result of deleting one occurrence of X from the list Xs.

```
select(X, [X | Xs], Xs).
select(X, [Y | Xs], [Y | Zs]) ← select(X, Xs, Zs).
```

As in the case of the APPEND program, it is most advantageous to use the level mapping

$$|\text{select}(xs, ys, zs)| = \min(|ys|, |zs|).$$

Then SELECT is recurrent w.r.t | | and if ys is a list or zs is a list, all SLD-derivations of SELECT ∪ { ← select(xs, ys, zs)} are finite.

2.3.4 Sum

Finally, consider the following program SUM:

```
sum(X, Y, Z) ←
     X, Y, Z are natural numbers such that Z is the sum of X and Y.
sum(X, s(Y), s(Z)) ← sum(X, Y, Z).
sum(X, 0, X).
```

Again, it is most advantageous to use here the level mapping

$$|\text{sum}(x, y, z)| = \min(\text{size}(y), \text{size}(z)),$$

where for a term t, size(t) denotes the number of symbols in t.

Then SUM is recurrent w.r.t. | | and for a ground y or z, sum(x, y, z) is bounded w.r.t. | |. By Corollary 2.6 SUM is terminating and by Corollary 2.5 for a ground y or z, all SLD-derivations of SUM ∪ { ← sum(x, y, z)} are finite.

3 Left termination

3.1 Motivation

Because of Corollary 2.11, recurrent programs and bounded goals are too restrictive concepts to deal with PROLOG programs, as a larger class of programs and goals is terminating when adopting a specific selection rule, e.g. PROLOG selection rule.

Example 2

(i) First we consider a terminating program P such that for some goal G all LD-derivations of $P \cup \{G\}$ are finite, whereas some SLD-derivation of $P \cup \{G\}$ is infinite.

Examine the following program EVEN:

```
even(X) ←
     X is an even natural number.

even(s(s(X))) ← even(X).
even(0).

lte(X, Y) ←
```

X, Y are natural numbers such that X is smaller than or equal to Y.

lte((s(X), s(Y)) ← lte(X, Y).
lte(0, Y).

EVEN is recurrent with
$$|\mathtt{even}(x)| = \mathrm{size}(x)$$
and
$$|\mathtt{lte}(x,y)| = \min(\mathrm{size}(x), \mathrm{size}(y))$$

so by Corollary 2.6 it is terminating. Now consider the goal:
$$G = \leftarrow \mathtt{lte}(X, s^{100}(0)), \mathtt{even}(X)$$

which is supposed to compute the even numbers not exceeding 100. One can show that all LD-derivations of G are finite, whereas there exists an infinite SLD-derivation when the rightmost selection rule is used. As a consequence of Corollary 2.5 the goal G is not bounded, although it can be evaluated by a finite PROLOG computation.

This example is a contrived instance of the *generate-and-test* programming technique. This technique involves two procedures, one which generates the set of candidates, and another which tests whether these candidates are solutions to the problem. Actually, most PROLOG programs that are implementations of the "generate-and-test" technique are not recurrent, as they heavily depend on the left-to-right order of evaluation, like the above goal.

(ii) Next, we consider a program P which is not terminating but such that all LD-derivations starting with a ground goal are finite. The following NAIVE REVERSE program is often used as a benchmark for PROLOG applications:

reverse(Xs, Ys) ← Ys is a reverse of the list Xs.

reverse([X | Xs], Ys) ←
    reverse(Xs, Zs),
    app(Zs, [X], Ys).
reverse([], []).

augmented by the APPEND program.

It is easy to check that the ground goal ← reverse(xs, ys), for a list xs with at least two elements and an arbitrary list ys has an infinite SLD-derivation, obtained by using the selection rule which selects the leftmost atom at the first two steps, and the second leftmost atom afterwards. Thus reverse is not terminating. However, one can show that all LD-derivations starting with a goal ← reverse(s, y) for s ground (or s list) are finite.

(iii) More generally, consider the following program DC, representing a (binary) *divide and conquer* schema; it is parametric w.r.t. the relations base, conquer, divide and merge.

```
dcsolve(X, Y) ←
    base(X),
    conquer(X, Y).
dcsolve(X, Y) ←
    divide(X, X0, X1, X2),
    dcsolve(X1, Y1),
    dcsolve(X2, Y2),
    merge(X0, Y1, Y2, Y).
```

Many programs naturally fit into this schema, or its generalization to non fixed arity of the relations divide/merge. Unfortunately, DC is not recurrent: it suffices to take a ground instance of the recursive clause with X=a, X1=a, Y=b, Y1=b, and observe that the atom dcsolve(a, b) occurs both in the head and in the body of such a clause. In this example, the leftmost selection rule is needed to guarantee that the input data is divided into subcomponents before recurring on such subcomponents.

To cope with these difficulties we first modify the definition of a terminating program.

3.2 Left terminating programs

Definition 3.1 *A program is called left terminating if all its LD-derivations starting with a ground goal are finite.*

This notion of termination is clearly more appropriate for the study of PROLOG programs than that of a terminating program. To prove that a program is left terminating, and to characterize the goals that terminate w.r.t. such a program, we introduce the following concepts due to Apt and Pedreschi (1993).

Definition 3.2 *Let P be a program, $|\ |$ a level mapping for P and I a (not necessarily Herbrand) interpretation of P.*

- *A clause of P is called acceptable w.r.t. $|\ |$ and I if I is its model and for every ground instance $A \leftarrow \mathbf{A}, B, \mathbf{B}$ of it such that $I \models \mathbf{A}$*

$$|A| > |B|.$$

In other words, for every ground instance $A \leftarrow B_1, \ldots, B_n$ of the clause

$$|A| > |B_i| \text{ for } i \in [1, \bar{n}],$$

where

$$\bar{n} = \min(\{n\} \cup \{i \in [1, n] \mid I \not\models B_i\}).$$

- A program P is called acceptable w.r.t. $|\ |$ and I if all its clauses are. P is called acceptable if it is acceptable w.r.t. some level mapping and an interpretation of P.

The use of the premise $I \models \mathbf{A}$ forms the *only* difference between the concepts of recurrence and acceptability. Intuitively, this premise expresses the fact that when in the evaluation of the goal $\leftarrow \mathbf{A}, B, \mathbf{B}$ using the leftmost selection rule the atom B is reached, the atoms $\mathbf{A}$ are already refuted. Consequently, by the soundness of the LD-resolution, these atoms are all true in I.

Alternatively, we may define $\bar{n}$ by

$$\bar{n} = \begin{cases} n & \text{if } I \models B_1 \wedge \cdots \wedge B_n, \\ i & \text{if } I \models B_1 \wedge \cdots \wedge B_{i-1} \text{ and } I \not\models B_1 \wedge \cdots \wedge B_i. \end{cases}$$

Thus, given a level mapping $|\ |$ for P and an interpretation I of P, in the definition of acceptability w.r.t. $|\ |$ and I, for every ground instance $A \leftarrow B_1, \ldots, B_n$ of a clause in P, we only require that the level of A is higher than the level of B_i's in a certain prefix of $B_1, \ldots, B_n$. Which B_i's are taken into account is determined by the model I. If $I \models B_1 \wedge \cdots \wedge B_n$ then all of them are considered and otherwise only those whose index is $\leq \bar{n}$, where $\bar{n}$ is the least index i for which $I \not\models B_i$.

The following observation shows that the notion of acceptability generalizes that of recurrence.

Lemma 3.3 *A program is recurrent w.r.t.* $|\ |$ *iff it is acceptable w.r.t.* $|\ |$ *and* B_P.

Our aim is to prove that the notions of acceptability and left termination coincide. To this end we need the notion of boundedness. The concept of a bounded goal used here differs from that introduced in Definition 2.3 in that it takes into account the interpretation I. This results in a more complicated definition.

In what follows, assume that the maximum function $max : 2^\omega \to N \cup \{\infty\}$ is defined as:

$$max\ S = \begin{cases} 0 & \text{if } S = \emptyset, \\ n & \text{if } S \text{ is finite and non-empty,} \\ & \text{and } n \text{ is the maximum of } S, \\ \infty & \text{if } S \text{ is infinite.} \end{cases}$$

Then $max\ S < \infty$ iff the set S is finite.

Definition 3.4 Let P be a program, $|\ |$ a level mapping for P and I an interpretation of P.

- With each goal $G = \leftarrow A_1,\ldots, A_n$ we associate n sets of natural numbers defined as follows, for $i \in [1, n]$:

$$|G|_i^I = \{|A'_i|\ |\ \leftarrow A'_1,\ldots, A'_n \text{ is a ground instance of } G \text{ and } I \models A'_1 \wedge \cdots \wedge A'_{i-1}\}.$$

- A goal G is called bounded w.r.t. $|\ |$ and I if $|G|_i^I$ is finite, for $i \in [1, n]$.
- For $G = \leftarrow A_1,\ldots, A_n$ bounded w.r.t. $|\ |$ and I we define a multiset $|G|_I$ of natural numbers as follows:

$$|G|_I = bag\,(max\,|G|_1^I,\ldots,\,max\,|G|_n^I).$$

- For G bounded w.r.t. $|\ |$ and I, and $k \geq 0$, we say that G is bounded by k (w.r.t. $|\ |$ and I) if $k \geq h$ for $h \in |G|_I$.

Note that a goal G is bounded w.r.t. $|\ |$ and B_P iff it is bounded w.r.t. $|\ |$ in the sense of Definition 2.3.

Lemma 3.5 Let P be a program that is acceptable w.r.t. a level mapping $|\ |$ and an interpretation I. Let G_1 be a goal that is bounded w.r.t. $|\ |$ and I, and let G_2 be an LD-resolvent of G_1 from P. Then

(i) G_2 is bounded w.r.t. $|\ |$ and I,

(ii) $|G_2|_I \prec_m |G_1|_I$.

Proof. It suffices to prove the following claims in which we refer to the given level mapping and interpretation I. ∎

Claim 1 An instance G' of a bounded goal $G = \leftarrow A_1,\ldots, A_n$ is bounded and $|G'|_I \preceq_m |G|_I$.

Proof. It suffices to note that $|G'|_i^I \subseteq |G|_i^I$ for $i \in [1, n]$. ∎

Claim 2 For every acceptable clause $A \leftarrow \mathbf{B}$ and sequence of atoms $\mathbf{C}$, if $\leftarrow A, \mathbf{C}$ is bounded, then $\leftarrow \mathbf{B}, \mathbf{C}$ is bounded and $|\leftarrow \mathbf{B}, \mathbf{C}|_I \prec_m |\leftarrow A, \mathbf{C}|_I$.

Proof. Let $\mathbf{B} = B_1,\ldots, B_n$ and $\mathbf{C} = C_1,\ldots, C_m$, for $n, m \geq 0$. We first prove the following facts.

Fact 1 For $i \in [1, n]$, $|\leftarrow B_1,\ldots, B_n, C_1,\ldots, C_m|_i^I$ is finite, and

$$max|\leftarrow B_1,\ldots, B_n, C_1,\ldots, C_m|_i^I < max|\leftarrow A, C_1,\ldots, C_m|_1^I.$$

Proof. We have

$$max|\leftarrow B_1,\ldots, B_n, C_1,\ldots, C_m|_i^I$$

$$\begin{aligned}= \quad & \{\text{Definition 3.4 }\}\\
& max\{|B'_i| \mid \; \leftarrow B'_1, \ldots, B'_n \text{ is a ground instance of } \leftarrow \mathbf{B}\\
& \text{ and } I \models B'_1 \wedge \cdots \wedge B'_{i-1}\}\\
= \quad & \{\text{for some } A',\; A' \leftarrow B'_1, \ldots, B'_n \text{ is a ground instance of } A \leftarrow \mathbf{B}\}\\
& max\{|B'_i| \mid A' \leftarrow B'_1, \ldots, B'_n \text{ is a ground instance of } A \leftarrow \mathbf{B}\\
& \text{ and } I \models B'_1 \wedge \cdots \wedge B'_{i-1}\}\\
< \quad & \{\text{Definition 3.2 and the fact that }\\
& \forall x \in S \; \exists y \in R : x < y \text{ implies } max\, S < max\, R\}\\
& max\{|A'| \mid A' \text{ is a ground instance of } A\}\\
= \quad & \{\text{Definition 3.4}\}\\
& max| \leftarrow A, C_1, \ldots, C_m|^I_1.
\end{aligned}$$

Fact 2 For $j \in [1,m]$, $| \leftarrow B_1, \ldots, B_n, C_1, \ldots, C_m|^I_{j+n}$ is finite, and

$$max| \leftarrow B_1, \ldots, B_n, C_1, \ldots, C_m|^I_{j+n} \leq max| \leftarrow A, C_1, \ldots, C_m|^I_{j+1}.$$

Proof. We have

$$\begin{aligned}& max| \leftarrow B_1, \ldots, B_n, C_1, \ldots, C_m|^I_{j+n}\\
= \quad & \{\text{Definition 3.4}\}\\
& max\{|C'_j| \mid \; \leftarrow B'_1, \ldots, B'_n, C'_1, \ldots, C'_m \text{ is a ground instance of}\\
& \leftarrow \mathbf{B}, \mathbf{C} \text{ and } I \models B'_1 \wedge \cdots \wedge B'_n \wedge C'_1 \wedge \cdots \wedge C'_{j-1}\}\\
\leq \quad & \{\text{for some } A',\; A' \leftarrow B'_1, \ldots, B'_n \text{ is a ground instance of } A \leftarrow \mathbf{B},\\
& I \text{ is a model of } P, \text{ and } S \subseteq R \text{ implies } max\, S \leq max\, R\}\\
& max\{|C'_j| \mid \; \leftarrow A', C'_1, \ldots, C'_m \text{ is a ground instance of } \leftarrow A, \mathbf{C}\\
& \text{ and } I \models A' \wedge C'_1 \wedge \cdots \wedge C'_{j-1}\}\\
= \quad & \{\text{Definition 3.4}\}\\
& max| \leftarrow A, C_1, \ldots, C_m|^I_{j+1}.
\end{aligned}$$

As a consequence of Facts 1 and 2 $\leftarrow \mathbf{B}, \mathbf{C}$ is bounded and

$$bag(max| \leftarrow \mathbf{B}, \mathbf{C}|^I_1, \ldots, max| \leftarrow \mathbf{B}, \mathbf{C}|^I_{n+m}) \prec_m$$
$$bag(max| \leftarrow A, \mathbf{C}|^I_1, \ldots, max| \leftarrow A, \mathbf{C}|^I_{m+1})$$

which establishes the claim. ∎

Corollary 3.6 *Let P be an acceptable program and G a bounded goal. Then all LD-derivations of $P \cup \{G\}$ are finite.*

Proof. The multiset ordering is well-founded. ∎

Corollary 3.7 *Every acceptable program is left terminating.*

Proof. Every ground goal is bounded. ∎

We now prove the converse of Corollary 3.7. To this end we proceed analogously as in the case of terminating programs and analyze the size of finite LD-trees. We need the following analogue of Lemma 2.9, where for a program P and a goal G we now denote by $nodes_P(G)$ the number of nodes in the LD-tree for $P \cup \{G\}$.

Lemma 3.8 *Let P be a program and G a goal such that the LD-tree for $P \cup \{G\}$ is finite. Then*

(i) *for all substitutions θ, $nodes_P(G\theta) \leq nodes_P(G)$,*

(ii) *for all prefixes H of G, $nodes_P(H) \leq nodes_P(G)$,*

(iii) *for all non-root nodes H in the LD-tree for $P \cup \{G\}$, $nodes_P(H) < nodes_P(G)$.*

Proof

(i) By the lifting lemma (see (Lloyd 1987)) we conclude that to every LD-derivation of $P \cup \{G\theta\}$ with input clauses $C_1, C_2, \ldots$, there corresponds an LD-derivation of $P \cup \{G\}$ with variants of input clauses $C_1, C_2, \ldots$, of the same or larger length. This implies the claim.

(ii) Consider a prefix $H = \leftarrow A_1, \ldots, A_k$ of $G = \leftarrow A_1, \ldots, A_n$ $(n \geq k)$. By an appropriate renaming of variables (formally justified by the variant lemma (see (Apt 1990)) we can assume that all input clauses used in the LD-tree for $P \cup \{H\}$ have no variables in common with G. We can now transform the LD-tree for $P \cup \{H\}$ into an initial subtree of the LD-tree for $P \cup \{G\}$ by replacing in it a node $\leftarrow \mathbf{B}$ by $\leftarrow \mathbf{B}, A_{k+1}\theta, \ldots, A_n\theta$, where θ is the composition of the mgu's used on the path from the root H to the node $\leftarrow \mathbf{B}$. This implies the claim.

(iii) Immediate by the definition. ∎

We can now demonstrate the desired result.

Theorem 3.9 *Let P be a left terminating program. Then for some level mapping $|\ |$ and an interpretation I of P*

(i) *P is acceptable w.r.t. $|\ |$ and I,*

(ii) *for every goal G, G is bounded w.r.t. $|\ |$ and I iff all LD-derivations of $P \cup \{G\}$ are finite.*

Proof. Define the level mapping by putting for $A \in B_P$

$$|A| = nodes_P(\leftarrow A).$$

Since P is left terminating, this level mapping is well defined. Next, choose

$$I = \{A \in B_P \mid \text{there is an LD-refutation of } P \cup \{\leftarrow A\}\}.$$

By the strong completeness of SLD-resolution, $I = M_P$, so I is a model of P.

First we prove one implication of (ii).

(ii1) Consider a goal G such that all LD-derivations of $P \cup \{G\}$ are finite. We prove that G is bounded by $nodes_P(G)$ w.r.t. $|\ |$ and I.

To this end take $\ell \in \cup |[G]|_I$. For some ground instance $\leftarrow A_1, \ldots, A_n$ of G and $i \in [1, \bar{n}]$, where

$$\bar{n} = \min(\{n\} \cup \{i \in [1,n] \mid I \not\models A_i\}),$$

we have $\ell = |A_i|$. We now calculate

$\quad nodes_P(G)$
$\geq \quad$ {Lemma 3.8 (i)}
$\quad nodes_P(\leftarrow A_1, \ldots, A_n)$
$\geq \quad$ {Lemma 3.8 (ii)}
$\quad nodes_P(\leftarrow A_1, \ldots, A_{\bar{n}})$
$\geq \quad$ {Lemma 3.8 (iii), noting that for $j \in [1, \bar{n}-1]$
$\quad\quad$ there is an LD-refutation of $P \cup \{\leftarrow A_1, \ldots, A_j\}$}
$\quad nodes_P(\leftarrow A_i, \ldots, A_{\bar{n}})$
$\geq \quad$ {Lemma 3.8 (ii)}
$\quad nodes_P(\leftarrow A_i)$
$= \quad$ {definition of $|\ |$}
$\quad |A_i|$
$= \quad \ell.$

(i) We now prove that P is acceptable w.r.t. $|\ |$ and I. Take a clause $A \leftarrow B_1, \ldots, B_n$ in P and its ground instance $A\theta \leftarrow B_1\theta, \ldots, B_n\theta$. We need to show that

$$|A\theta| > |B_i\theta| \text{ for } i \in [1, \bar{n}],$$

where

$$\bar{n} = \min(\{n\} \cup \{i \in [1,n] \mid I \not\models B_i\theta\}).$$

We have $A\theta\theta \equiv A\theta$, so $A\theta$ and A unify. Let $\mu = \text{mgu}(A\theta, A)$. Then $\theta = \mu\delta$ for some δ. By the definition of LD-resolution, $\leftarrow B_1\mu, \ldots, B_n\mu$ is an LD-resolvent of $\leftarrow A\theta$.

Then for $i \in [1, \bar{n}]$

$\quad |A\theta|$
$= \quad$ {definition of $|\ |$}
$\quad nodes_P(\leftarrow A\theta)$

$\qquad >\qquad$ {Lemma 3.8 (iii), $\leftarrow B_1\mu, \ldots, B_n\mu$ is a resolvent of $\leftarrow A\theta$}
$\qquad\qquad nodes_P\ (\leftarrow B_1\mu, \ldots, B_n\mu)$
$\qquad \geq\qquad$ {part (ii1), with $G := \leftarrow B_1\mu, \ldots, B_n\mu$ and $A_i := B_i\theta$}
$\qquad |B_i\theta|$.

(ii2) Consider a goal G which is bounded w.r.t. $|\ |$ and I. Then by (i) and Corollary 3.6 all LD-derivations of $P \cup \{G\}$ are finite. ∎

Corollary 3.10 *A program is left terminating iff it is acceptable.*

Proof. By Corollary 3.7 and Theorem 3.9. ∎

3.3 Examples

The equivalence between the left terminating and acceptable programs provides us with a method of proving termination of PROLOG programs. The level mapping and the model used in the proof of Theorem 3.9 were quite involved and relied on elaborate information about the program at hand, which is usually not readily available. However, in practical situations much simpler constructions suffice. We illustrate it by means of two examples. In these we use the previously defined function listsize $|\ |$, which assigns natural numbers to ground terms.

In the following, we present the proof of acceptability (w.r.t. a level mapping $|\ |$ and an interpretation I) of a given clause $C = A_0 \leftarrow A_1, \ldots, A_n$ by means of the following *proof outline*:

$$\begin{array}{lll}
\{f_0\} & & \\
A_0 & \leftarrow & \{t_0\} \\
& A_1, & \{t_1\} \\
& \{f_1\} & \\
& \vdots & \\
& A_{n-1}, & \{t_{n-1}\} \\
& \{f_{n-1}\} & \\
& A_n. & \{t_n\} \\
& \{f_n\} &
\end{array}$$

Here, t_i and f_i, for $i \in [0, n]$ are integer expressions and first order formulas, respectively, such that all ground instances of the following properties are satisfied:

(1) $t_i = |A_i|$, for $i \in [0, n]$,
(2) $f_i \equiv I \models A_i$, for $i \in [0, n]$,
(3) $f_1 \wedge \cdots \wedge f_n \Rightarrow f_0$,

(4) For $i \in [1, n]$: $f_1 \wedge \cdots \wedge f_{i-1} \Rightarrow t_0 > t_i$.

We omit $\{f_i\}$ (resp. $\{t_i\}$) in the proof outlines if $f_i = $ **true** (resp. $t_i = 0$.) It is immediate that a proof outline satisfying properties 1 to 4 corresponds to the proofs that I is a model of the clause C, and that C is acceptable w.r.t. $| \ |$ and I. We found it convenient to use proof outlines to present the proofs of acceptability, as most steps in these proofs are trivial and can be omitted without loss of information.

3.3.1 Permutation

Consider the following program PERMUTATION:

```
perm(Xs, Ys) ←  Ys is a permutation of the list Xs.

perm(Xs, [X | Ys]) ←
    app(X1s, [X | X2s], Xs),
    app(X1s, X2s, Zs),
    perm(Zs, Ys).
perm([], []).
```

augmented by the APPEND program.

The intention is to invoke perm with its first argument instantiated. The first clause takes care of a non-empty list xs. One should first split it into two sublists x1s and [x | x2s] and concatenate x1s and x2s to get zs. If now ys is a permutation of zs, then [x | ys] is a permutation of xs. The second clause states that the empty list is a permutation of itself.

Observe the following:

- PERMUTATION is not recurrent. Indeed, consider the SLD-derivation of PERMUTATION $\cup$ { ← perm(xs, [x | ys])} with xs, x, ys ground, in whose second goal the middle atom app(x1s, x2s, zs) is selected. By repeatedly applying the recursive clause of APPEND we obtain an infinite derivation. Thus PERMUTATION is not terminating and so by Corollary 2.6 it is not recurrent.

- The Herbrand interpretation

$$I_{APP} = \{\text{app}(\text{xs}, \text{ys}, \text{zs}) \mid |\text{xs}| + |\text{ys}| = |\text{zs}|\}$$

is a model of the program APPEND. Indeed, I_{APP} is trivially a model of the non-recursive clause of the app relation and the following proof outline shows that I_{APP} is a model of the recursive clause:

$$\{1 + |\text{xs}| + |\text{ys}| = 1 + |\text{zs}|\}$$
$$\text{app}([x|\text{xs}], \text{ys}, [x|\text{zs}]) \qquad \leftarrow$$
$$\text{app}(\text{xs}, \text{ys}, \text{zs}).$$
$$\{|\text{xs}| + |\text{ys}| = |\text{zs}|\}$$

- The program PERMUTATION is acceptable w.r.t. the level mapping $|\ |$ and the interpretation I_{PERM} defined by

$$|perm(xs, ys)| = |xs| + 1,$$
$$|app(xs, ys, zs)| = \min(|xs|, |zs|),$$

$$I_{PERM} = [perm(Xs, Ys)] \cup I_{APP}.$$

Recall that $[A]$ for an atom A stands for the set of all ground instances of A. We already noted in Example 1 that APPEND is recurrent w.r.t. $|\ |$. The proof outline for the non-recursive clause of the perm relation is obvious. For the recursive clause take the following proof outline:

$$\text{perm}(xs, [x|ys]) \leftarrow \qquad\qquad \{|xs|+1\}$$
$$\text{app}(x1s, [x|x2s], xs), \qquad \{\min(|x1s|, |xs|)\}$$
$$\{|x1s|+1+|x2s| = |xs|\}$$
$$\text{app}(x1s, x2s, zs), \qquad \{\min(|x1s|, |zs|)\}$$
$$\{|x1s|+|x2s| = |zs|\}$$
$$\text{perm}(zs, ys). \qquad\qquad \{|zs|+1\}$$

Using Corollary 3.7 we conclude that PERMUTATION is acceptable. Moreover, we obtain that, for a list s, the atom perm(s,t) is rigid and hence bounded. Consequently, by Corollary 3.6, all LD-derivations of PERMUTATION $\cup \{\leftarrow \text{perm}(s,t)\}$ are finite.

3.3.2 Quicksort

Consider now the following program QUICKSORT:

qs(Xs, Ys) ← Ys is an ordered permutation of the list Xs.

```
qs([X | Xs], Ys) ←
    part(X, Xs, Littles, Bigs),
    qs(Littles, Ls),
    qs(Bigs, Bs),
    app(Ls, [X | Bs], Ys).
qs([], []).

part(X, [Y | Xs], [Y | Ls], Bs) ←
    X > Y, part(X, Xs, Ls, Bs).
part(X, [Y | Xs], Ls, [Y | Bs]) ←
    X ≤ Y, part(X, Xs, Ls, Bs).
part(X, [], [], []).
```

augmented by the APPEND program.

According to this sorting procedure, using its first element X, a list is first partitioned in two sublists, one consisting of elements smaller than X, and the other consisting of elements larger or equal than X. Then each sublist is quicksorted, and the resulting sorted sublists are appended with the element X put in the middle.

We assume that QUICKSORT operates on the domain of natural numbers over which the built-in relations $>$ and $\leq$, written in infix notation, are defined. We thus assume that this domain is part of the Herbrand universe of QUICKSORT.

Observe the following:

- QUICKSORT is not recurrent. In fact, consider the first clause instantiated with the grounding substitution

$$\{X/a,\ Xs/b,\ Ys/c,\ Littles/[a \mid b],\ Ls/c\}.$$

Then the ground atom qs([a | b], c) appears both in the head and the body of the resulting clause.
- The clauses defining the relation part are trivially recurrent with $|\text{part}(x, xs, ls, bs)| = |xs|$, $|s > t| = 0$ and $|s \leq t| = 0$.
- Extend now the above level mapping with

$$|\text{qs}(xs, ys)| = |xs|,$$
$$|\text{app}(xs, ys, zs)| = |xs|.$$

Recall that APPEND is recurrent w.r.t. $|\ |$. Next, define a Herbrand interpretation of QUICKSORT by putting

$$\begin{aligned}
I = & \ \{\text{qs}(xs, ys) \mid |xs| \geq |ys|\} \\
\cup & \ \{\text{part}(x, xs, ls, bs) \mid |xs| \geq |ls| + |bs|\} \\
\cup & \ \{\text{app}(xs, ys, zs) \mid |xs| + |ys| \geq |zs|\} \\
\cup & \ [X > Y] \\
\cup & \ [X \leq Y].
\end{aligned}$$

The following proof outlines show that QUICKSORT is acceptable w.r.t. $|\ |$ and I. The proof outlines for the non-recursive clauses are obvious and omitted.

$$\{1 + |xs| + |ys| \geq 1 + |zs|\}$$
$$\text{app}([x|xs], ys, [x|zs]) \quad \leftarrow$$
$$\text{app}(xs, ys, zs).$$
$$\{|xs| + |ys| \geq |zs|\}$$

$\{1 + |\text{xs}| \geq 1 + |\text{ls}| + |\text{bs}|\}$
part(x, [y|xs], [y|ls], bs) ←

$\qquad\qquad\qquad\qquad$ X > Y,

$\qquad\qquad\qquad\qquad$ part(x, xs, ls, bs).

$\qquad\qquad\qquad\qquad$ $\{|\text{xs}| \geq |\text{ls}| + |\text{bs}|\}$

$\{1 + |\text{xs}| \geq |\text{ls}| + 1 + |\text{bs}|\}$
part(x, [y|xs], ls, [y|bs]) ←

$\qquad\qquad\qquad\qquad$ X $\leq$ Y,

$\qquad\qquad\qquad\qquad$ part(x, xs, ls, bs).

$\qquad\qquad\qquad\qquad$ $\{|\text{xs}| \geq |\text{ls}| + |\text{bs}|\}$

$\{1 + |\text{xs}| \geq |\text{ys}|\}$
qs([x|xs], ys) ← $\qquad\qquad\qquad\qquad\qquad\qquad$ $\{1 + |\text{xs}|\}$

$\qquad\qquad$ part(x, xs, littles, bigs), $\qquad$ $\{|\text{xs}|\}$

$\qquad\qquad$ $\{|\text{xs}| \geq |\text{littles}| + |\text{bigs}|\}$

$\qquad\qquad$ qs(littles, ls), $\qquad\qquad\qquad$ $\{|\text{littles}|\}$

$\qquad\qquad$ $\{|\text{littles}| \geq |\text{ls}|\}$

$\qquad\qquad$ qs(bigs, bs), $\qquad\qquad\qquad\quad$ $\{|\text{bigs}|\}$

$\qquad\qquad$ $\{|\text{bigs}| \geq |\text{bs}|\}$

$\qquad\qquad$ app(ls, [x|bs], ys). $\qquad\qquad\quad$ $\{|\text{ls}|\}$

$\qquad\qquad$ $\{|\text{ls}| + 1 + |\text{bs}| \geq |\text{ys}|\}$

Using Corollary 3.7 we conclude that QUICKSORT is acceptable. Moreover, we obtain that, for a list s, the atom qs(s,t) is rigid and hence bounded. By Corollary 3.6 we conclude that all LD-derivations of QUICKSORT ∪{ ← qs(s, t)} are finite.

4 A modular approach to termination
4.1 Drawbacks of the proof method
The proof method for (left) termination introduced in the previous sections suffers from two drawbacks.

- The level mapping used in the proof of recurrence/acceptability is sometimes different from the expected natural candidate. Consider for instance the program PERMUTATION. The relation perm is defined by induction on the length of its first argument, which is a list, and therefore a natural candidate for |perm(xs,ys)| is |xs|. Nevertheless, it is needed to add 1 to such a value in order to enforce a strict decreasing from the relation perm to the relation app, as required by the definition of acceptability.
- The proposed proof method does not provide means for constructing modular proofs, hence no straightforward technique is available to combine proofs for separate programs to obtain proofs of combined programs.

As the module hierarchy of a program becomes more complex, such adjustments of natural level mappings become more artificial and consequently more difficult to discover. For example, to prove that the program

```
overlap(Xs, Ys)  ← member(X, Xs), member(X, Ys).
has_a_or_b(Xs)  ← overlap(Xs, [a,b]).
```

augmented by the MEMBER program.

is recurrent we need to put |overlap(xs, ys)| = |xs|+1 to enforce the decrease in the first clause and then |has_a_or_b(xs)| = |xs|+2 to enforce the decrease in the second clause.

Both drawbacks pose serious limitations for the practical applicability of the proposed proof method. First, as argued by De Schreye, Verschaetse and Bruynooghe (1992), unnatural level mappings are difficult to discover by automated tools. Secondly, modularity is essential in mastering the complexity of large-scaled programs. These drawbacks share their originating cause. The notions of recurrence/acceptability are based on the fact that level mappings decrease from clause heads to clause bodies. This is used for two different purposes:

(1) in (mutually) recursive calls, to ensure termination of (mutually) recursive procedures, and
(2) in non (mutually) recursive calls, to ensure that non (mutually) recursive procedures are called with terminating goals.

Although a decreasing of the level mappings is apparently essential for the first purpose, this is not the case for the second purpose, since a weaker condition can be adopted to ensure that non-recursive procedures are properly called.

The next subsections will elaborate on this idea, by presenting alternative definitions of recurrence/acceptability, which we qualify with the prefix *semi*. These notions are actually proved equivalent to the original ones, but they give rise to more flexible proof methods, which avoid the cited drawbacks.

4.2 Semi-recurrent programs

Following the intuition that recursive and non-recursive procedures should be handled separately in proving termination, we introduce a natural ordering over the relation names occurring in a program P, with the intention that for relations p and q, $p \sqsupseteq q$ holds if procedure p *can call* procedure q. The next definition makes this concept precise. We denote by Π_P the set of relations occurring in a program P.

Definition 4.1 *Let P be a program and p, q relations in Π_P.*

(i) *We say that p refers to q in P if there is a clause in P that uses p in its head and q in its body.*

(ii) *We say that p depends on q in P, and write $p \sqsupseteq q$, if (p, q) is in the reflexive, transitive closure of the relation refers to.*

Observe that according to the above definition $p \simeq q \equiv p \sqsupseteq q \wedge q \sqsupseteq p$ means that p and q are mutually recursive (i.e., they are in the same recursive clique), and $p \sqsupset q \equiv p \sqsupseteq q \wedge q \not\sqsupseteq p$ means that p calls q as a subprogram. It is important to notice that the ordering $\sqsupset$ over Π_P is well founded.

The following definition of *semi-recurrence* exploits the ordering over the relation names. The level mapping is required to decrease from an atom A in the head of a clause to an atom B in the body of that clause only if the relations of A and B are mutually recursive. Additionally, the level mapping is required not to increase from A to B if the relations of A and B are not mutually recursive. We adopt the notation $rel(A)$ to denote the relation symbol occurring in atom A.

Definition 4.2

- *A clause is called semi-recurrent w.r.t. a level mapping $|\ |$, if for every ground instance $A \leftarrow \mathbf{A}, B, \mathbf{B}$ of it*
  (i) $|A| > |B|$ *if $rel(A) \simeq rel(B)$,*
  (ii) $|A| \geq |B|$ *if $rel(A) \sqsupset rel(B)$.*
- *A program P is called semi-recurrent w.r.t. a level mapping $|\ |$ if all its clauses are. P is called semi-recurrent if it is semi-recurrent w.r.t. some level mapping.*

The following observation is immediate.

Lemma 4.3 *If a program is recurrent w.r.t. $|\ |$, then it is semi-recurrent w.r.t. $|\ |$.*

The converse of Lemma 4.3 also holds, in a sense made precise by the following result.

Lemma 4.4 *If a program is semi-recurrent w.r.t. $|\ |$, then it is recurrent w.r.t. a level mapping $\|\ \|$. Moreover, for each atom A, if A is bounded* w.r.t. $|\ |$, then A is bounded w.r.t. $||\ ||$.

Proof. In order to define the level mapping $||\ ||$, we first introduce (by overloading the symbol $|\ |$) a mapping $|\ |: \Pi_P \to N$ such that, for $p, q \in \Pi_P$:

$$p \simeq q \text{ implies } |p| = |q|, \quad (4.1)$$
$$p \sqsupset q \text{ implies } |p| > |q|. \quad (4.2)$$

A mapping $|\ |$ satisfying properties (4.1) and (4.2) obviously exists, as Π_P is finite. Note that this mapping preserves the $\sqsupset$ ordering. Next, we define a level mapping $||\ ||$ for P by putting for $A \in B_P$:

$$||A|| = |A| + |rel(A)| \quad (4.3)$$

We now prove that P is recurrent w.r.t. $||\ ||$. Let $A \leftarrow \mathbf{A}, B, \mathbf{B}$ be a ground instance of a clause from P. The following two cases arise:

Case 1 $rel(A) \simeq rel(B)$.
We calculate:

$$
\begin{aligned}
& ||A|| \\
= \quad & \{(4.3)\} \\
& |A| + |rel(A)| \\
> \quad & \{|A| > |B| \text{ by Definition 4.2 (i)}\} \\
& |B| + |rel(A)| \\
= \quad & \{|rel(A)| = |rel(B)| \text{ by } rel(A) \simeq rel(B) \text{ and } (4.1)\} \\
& |B| + |rel(B)| \\
= \quad & \{(4.3)\} \\
& ||B||.
\end{aligned}
$$

Case 2 $rel(A) \sqsupset rel(B)$.
We calculate:

$$
\begin{aligned}
& ||A|| \\
= \quad & \{(4.3)\} \\
& |A| + |rel(A)| \\
> \quad & \{|rel(A)| > |rel(B)| \text{ by } rel(A) \sqsupset rel(B) \text{ and } (4.2)\} \\
& |A| + |rel(B)| \\
\geq \quad & \{|A| \geq |B| \text{ by } rel(A) \sqsupset rel(B) \text{ and Definition 4.2 (ii)}\} \\
& |B| + |rel(B)| \\
= \quad & \{(4.3)\} \\
& ||B||.
\end{aligned}
$$

In both cases we proved $||A|| > ||B||$, which establishes the first claim. The second claim follows directly from the definition of $|| \ ||$. ∎

The following is an immediate conclusion of Lemmata 4.3 and 4.4.

Corollary 4.5 *A program is recurrent iff it is semi-recurrent.*

In what follows we study conditions which allow us to deduce termination of a program from termination of its components. The simplest form of program composition takes place when a program is constructed from two subprograms which use disjoint sets of relations. The following obvious composition theorem allows us to deal with this case.

Theorem 4.6 *Let P and Q be two programs such that no relation occurs in both of them. Suppose that*

- *Q is semi-recurrent w.r.t. level mapping $|\ |_Q$,*
- *P is semi-recurrent w.r.t. level mapping $|\ |_P$.*

Then $P \cup Q$ is semi-recurrent w.r.t. $|\ |$ defined as follows:

$$|A| = \begin{cases} |A|_P & \text{if } rel(A) \text{ is defined in } P, \\ |A|_Q & \text{if } rel(A) \text{ is defined in } Q. \end{cases}$$

Obviously, this theorem is of very limited use. We now consider a situation when a program uses another one as a subprogram. The following notion of *extension* of a program formalizes this situation.

Definition 4.7 *Let P and Q be two programs.*

(i) A relation p is defined in a program P if p occurs in the head of a clause from P.

(ii) P extends Q if no relation defined in P occurs in Q.

Informally, P extends Q if P defines new (w.r.t. Q) relations. From now we assume without loss of generality that, for a given program P and a level mapping $|\ |$ for P, $|A| = 0$ if $rel(A)$ is not defined in P. Notice that such an assumption is indeed immaterial for the notion of (semi-)recurrence, since if $rel(A)$ does not occur in the head of any clause of P, then any constraint put on $|A|$ is satisfied when $|A| = 0$.

Observe that the definition of semi-recurrence allows us to compose termination proofs. Indeed, the following result holds.

Theorem 4.8 *Let P and Q be two programs such that P extends Q. Suppose that*

(1) Q is semi-recurrent w.r.t. $|\ |_Q$,

(2) P is semi-recurrent w.r.t. $|\ |_P$,

(3) for every ground instance $A \leftarrow \mathbf{A}, B, \mathbf{B}$ of a clause of P

$$|A|_P \geq |B|_Q \text{ if } rel(B) \text{ is defined in } Q.$$

Then $P \cup Q$ is semi-recurrent w.r.t. $|\ |$ defined as follows:

$$|A| = \begin{cases} |A|_P & \text{if } rel(A) \text{ is defined in } P, \\ |A|_Q & \text{if } rel(A) \text{ is defined in } Q. \end{cases} \quad (4.4)$$

Proof. It suffices to note that for every ground instance $A \leftarrow \mathbf{A}, B, \mathbf{B}$ of a clause from $P \cup Q$ the following two implications hold:
(i) if $rel(A) \simeq rel(B)$, then either both relations are defined in P or both are defined in Q,
(ii) if $rel(A) \sqsupset rel(B)$, then either $rel(A)$ is not defined in P. ∎

This theorem suggests a natural way of composing termination proofs *provided* the level mappings of the programs P and Q satisfy condition 3. In general, it is difficult to expect that two independently constructed level mappings happen to satisfy such a relation. (An example illustrating this complication can be found below.)

Consequently, we need a more general approach. The result we now present makes it possible to construct termination proofs in a modular way in full generality and is the main motivation for the introduction of the notion of semi-recurrence.

Theorem 4.9 *Let P and Q be two programs such that P extends Q. Suppose that*

(1) *Q is semi-recurrent w.r.t. $|\ |_Q$,*
(2) *P is semi-recurrent w.r.t. $|\ |_P$,*
(3) *there exists a level mapping $||\ ||_P$ such that for every ground instance $A \leftarrow \mathbf{A}, B, \mathbf{B}$ of a clause from P*
   (a) $||A||_P \geq ||B||_P$ *if $rel(B)$ is defined in P,*
   (b) $||A||_P \geq |B|_Q$ *if $rel(B)$ is defined in Q.*

Then $P \cup Q$ is semi-recurrent w.r.t. $|\ |$ defined as follows:

$$|A| = \begin{cases} |A|_P + ||A||_P & \text{if } rel(A) \text{ is defined in } P, \\ |A|_Q & \text{if } rel(A) \text{ is defined in } Q. \end{cases} \quad (4.5)$$

Proof. It suffices to prove that each clause from P is semi-recurrent w.r.t. $|\ |$. Let $A \leftarrow \mathbf{A}, B, \mathbf{B}$ be a ground instance of a clause from P. The following two cases arise:
Case 1 $rel(A) \simeq rel(B)$.
Then by Definition 4.7, $rel(B)$ is defined in P. According to Definition 4.2(i) we need to prove $|A| > |B|$. We calculate:

$|A|$
$= \quad \{(4.4)\}$

$$|A|_P + ||A||_P$$
$> \quad \{|A|_P > |B|_P \text{ by assumption 2 and Definition 4.2 (i)},$
$\quad\quad \text{and } ||A||_P > ||B||_P \text{ by assumption 3(a)}\}$
$$|B|_P + ||B||_P$$
$= \quad \{(4.4)\}$
$$|B|.$$

Case 2 $rel(A) \sqsupset rel(B)$.
According to Definition 4.2(ii), we need to prove $|A| \geq |B|$. Two subcases arise:

Subcase 2.1 $rel(B)$ is defined in P.
We calculate:

$$|A|$$
$= \quad \{(4.4)\}$
$$|A|_P + ||A||_P$$
$\geq \quad \{|A|_P \geq |B|_P \text{ by assumption 2 and Definition 4.2 (ii)},$
$\quad\quad \text{and } ||A||_P \geq ||B||_P \text{ by assumption 3(a)}\}$
$$|B|_P + ||B||_P$$
$= \quad \{(4.4)\}$
$$|B|.$$

Subcase 2.2 $rel(B)$ is defined in Q.
We calculate:

$$|A|$$
$= \quad \{(4.4)\}$
$$|A|_P + ||A||_P$$
$\geq \quad \{||A||_P \geq |B|_Q \text{ by assumption 3(b)}\}$
$$|A|_P + |B|_Q$$
$\geq \quad |B|_Q$
$= \quad \{(4.4)\}$
$$|B|.$$

∎

4.3 Methodology

Theorems 4.6, 4.8 and 4.9 provide us with an incremental, bottom-up method for proving termination of logic programs. Given a program P, the method can be informally illustrated as follows.

(1) Partition the relation names in P in the equivalence classes w.r.t. the equivalence $\simeq$ induced by the "depends on" relation $\sqsupseteq$. Such equivalence classes correspond to the recursive cliques of program P. Let $P_1, \ldots, P_n$ be the partition of the clauses from P such that each P_i contains the clauses defining the relation(s) belonging to the same equivalence class. The relation $\sqsupseteq$ defined on the relations induces a corresponding well founded ordering $>$ on the programs P_i:

$P_i > P_j$ iff $p \sqsupseteq q$ for some p defined in P_i and q defined in P_j.

(2) Prove by induction w.r.t. the ordering $>$ that for every program P_i, $i \in [1, n]$ the program $P_i \cup \bigcup_{P_j < P_i} P_j$ is semi-recurrent.

The base case. Consider all P_i, $i \in [1, n]$, which are minimal w.r.t. $>$.

- Prove that each such P_i is semi-recurrent (w.r.t. some $|\ |_{P_i}$).
  Notice that this is the same as proving that P_i is recurrent w.r.t. $|\ |_{P_i}$, as procedures in P_i do not call any subprograms.

The induction step. Consider a P_i, $i \in [1, n]$, such that all P_j for which $P_j < P_i$ have already been proved semi-recurrent.

- Prove that P_i (in isolation) is recurrent w.r.t. some $|\ |_{P_i}$.
  Notice that the assumption that $|A|_{P_i} = 0$ if $rel(A)$ is not defined in P_i allows us to abstract from the relations that are not defined in P_i. Consequently, we only need to prove that $|\ |_{P_i}$ decreases on (mutually) recursive calls. This facilitates the choice of a "natural" candidate for
  $|\ |_{P_i}$, which directly mirrors the inductive structure of the procedures defined in P_i.
- Use Theorem 4.6 to conclude that $\bigcup_{P_j < P_i} P_j$ is semi-recurrent.
- Use Theorem 4.8 or Theorem 4.9 to prove that $P_i \cup \bigcup_{P_j < P_i} P_j$ is semi-recurrent.
  Here we only need to come up with a level mapping $\|\ \|$ which is usually directly suggested by the level mappings $|\ |_{P_j}$, where $P_j < P_i$.

4.4 Examples

4.4.1 *Mergesort*

Consider the following program MERGESORT which is an instance of the divide and conquer schema:

```
ms(Xs, Ys) ←
    Ys is an ordered permutation of the list Xs.

ms([X, Y | Xs], Ys) ←
    split([X, Y | Xs], X1s, X2s),
    ms(X1s, Y1s),
```

```
    ms(X2s, Y2s),
    merge(Y1s, Y2s, Ys).
ms([X], [X]).
ms([], []).

split([X | Xs], [X | Ys], Zs) ←
    split(Xs, Zs, Ys).
split([], [], []).

merge([X | Xs], [Y | Ys], [X | Zs]) ←
    X ≤ Y,
    merge(Xs,[Y | Ys], Zs).
merge([X | Xs], [Y | Ys], [Y | Zs]) ←
    X > Y,
    merge([X | Xs], Ys, Zs).
merge([], Xs, Xs).
merge(Xs, [], Xs).
```

According to this sorting procedure, a list of length at least 2 is first split into two lists of roughly equal length (by means of the reversed order of parameters in the recursive call of split), then each sublist is mergesorted, and finally the resulting sorted sublists are merged, preserving the ordering.

Note that MERGESORT is not recurrent. Indeed, due to the introduction of the local variables X1,X2,Y1, Y2 in the body of the recursive clause defining ms, it is not terminating. By adding an additional parameter Bezem (1993) modified this program so that it becomes terminating:

```
ms(Xs, Ys, Xs) ←
    Ys is an ordered permutation of the list Xs.

ms([X, Y | Xs], Ys, [H | Ls]) ←
    split([X, Y | Xs], X1s, X2s, [H | Ls]),
    ms(X1s, Y1s, Ls),
    ms(X2s, Y2s, Ls),
    merge(Y1s, Y2s, Ys, [H | Ls]).
ms([X], [X], Ls).
ms([], [], Ls).

split([X | Xs], [X | Ys], Zs, [H | Ls]) ←
    split(Xs, Zs, Ys, Ls).
split([], [], [], Ls).

merge([X | Xs], [Y | Ys], [X | Zs], [H | Ls]) ←
    X ≤ Y,
    merge(Xs,[Y | Ys], Zs, Ls).
merge([X | Xs], [Y | Ys], [Y | Zs], [H | Ls]) ←
    X > Y,
    merge([X | Xs], Ys, Zs, Ls).
```

```
merge([], Xs, Xs, Ls).
merge(Xs, [], Xs, Ls).
```

(A misprint crept in to (Bezem 1993) where instead of calling `merge`, `ms` calls itself.) We prove this fact using Theorems 4.6 and 4.8. Call the above program MERGESORT' and denote the subprograms of MERGESORT' which define the relations `ms`, `split` and `merge` by MS, SPLIT and MERGE, correspondingly. Thanks to the addition of the last argument MS is recurrent w.r.t. the level mapping

$$|\mathtt{ms(xs,ys,ls)}| = |\mathtt{ls}|,$$

SPLIT is recurrent w.r.t. the level mapping

$$|\mathtt{split(xs,ys,zs,ls)}| = |\mathtt{ls}|,$$

and MERGE is recurrent w.r.t. the level mapping

$$|\mathtt{merge(xs,ys,zs,ls)}| = |\mathtt{ls}|.$$

By Theorem 4.6 SPLIT ∪ MERGE is recurrent w.r.t. | |. Assumption 3 of Theorem 4.8 applied to the programs MS and SPLIT ∪ MERGE is obviously satisfied, so we conclude by this theorem that MERGESORT' is semi-recurrent w.r.t. | |, and hence terminating.

To prove this fact Bezem (1993) used the concept of a recurrent program, which, to deal with the subprogram calls in the recursive clause defining `ms`, requires a more artificial level mapping in which |ms(xs, ys, ls)| = |ls|+1.

4.4.2 *Curry's type assignment*

Consider the following program for Curry's type assignment (see e.g. (Reddy 1986)). In Curry's type system, a *type assignment* E ⊢ M : T expresses the fact that λ-term M is assigned type T w.r.t. environment E. Here, λ-terms are represented using the function symbols `var` (for variables), `apply` (for application), and `lambda` (for λ-abstraction). Type terms are represented using the function symbol `arrow` (for the function type). For the sake of concreteness, we augment the program with extra constants (say v, w, z) representing λ-variables, and others (say Nat, Bool) representing basic types. Finally, environments are represented as lists of pairs (λ-variable, type term).

```
       type(E,M,T)  ←
          E ⊢ M : T
(t₁)   type(E,var(X),T)  ← in(E,X,T).
(t₂)   type(E,apply(M,N),T)  ←
          type(E,M,arrow(S,T)), type(E,N,S).
(t₃)   type(E,lambda(X,M),arrow(S,T))  ←
```

```
         type([(X,S)|E],M,T).

     in(E,X,T) ←
         X is bound to T in E
($i_1$)  in([(X,T)|E],X,T).
($i_2$)  in([(Y,T)|E],X,T) ← X ≠ Y, in(E,X,T).
```

Denote by CURRY the program formed by clauses t_1, t_2 and t_3, and by ENV the program formed by clauses i_1 and i_2. Clearly, CURRY extends ENV, and type⊐ in ⊐ ≠ in curry ∪ env. Observe the following:

- relation in is defined by induction on the length of its first argument, which is a list. As a result, the program ENV is recurrent w.r.t. $|\ |_{ENV}$ defined as:

$$|\text{in}(e, x, t)|_{ENV} = |e|.$$

- Relation type is defined by induction on the size of its first argument, which is a λ-term. As a result, the program CURRY is recurrent w.r.t. $|\ |_{CURRY}$ defined as:

$$|\text{type}(e, m, t)|_{CURRY} = \text{size}(m).$$

- In any derivation starting from a goal ← type(e, m, t), the length of the environment is bounded by $|e|+\text{size}(m)$, since the length of the environment is incremented together with the decrease of the size of the λ-term (clause t_3). As a result, by defining

$$\|\text{type}(e, m, t)\|_{CURRY} = |e| + \text{size}(m)$$

we satisfy for $\|\ \|_{CURRY}$ the assumptions 3(a) and (b) of Theorem 4.9.

Note that the level mappings $|\ |_{ENV}$ and $|\ |_{CURRY}$ do not satisfy condition 3 of Theorem 4.8, so this theorem cannot be used here.

As a consequence, by Theorem 4.9, Lemma 4.4 and Corollary 2.6 we conclude that CURRY ∪ ENV is terminating. Additionally, we obtain that a goal ← type(e, m, t) is bounded if e is a list and m is ground. This latter result is relevant, since it justifies the fact that program CURRY ∪ ENV can be used to implement type *inference* by means of the goals of the kind ← type(e, m, T), where e is a list, m is a ground λ-term, and T is a variable.

As a final remark, notice that it is possible to arrive at the same conclusion by showing directly that CURRY ∪ ENV is recurrent w.r.t. the level mapping $|\text{type}(e, m, t)| = |e| + 2 \times \text{size}(m)$, but such a level mapping is unnatural. Moreover, such a proof cannot be readily explained in a compositional way, as a combination of the separate proofs for CURRY and ENV.

4.4.3 *A relational* MAP *program*

Consider the following program MAP, implementing a relational equivalent of the ubiquitous higher-order combinator *map* of functional programming:

```
map([X1, ..., Xn], [Y1, ..., Yn]) ←
    p(Xi, Yi) holds for i ∈ [1,n].
map([X|Xs], [Y|Ys]) ←
    p(X, Y), map(Xs, Ys).
map([], []).
```

The program MAP is parametric w.r.t. relation p. Let P be a program defining the relation p, such that MAP extends P (hence: map $\sqsupset$ p.) Assume that P is recurrent w.r.t. $|\ |_P$ defined as $|p(x,y)|_P = f(x)$, where $f(x)$ denotes some function assigning natural numbers to ground terms.

We observe the following:

- The program MAP is trivially recurrent w.r.t. $|\ |_{MAP}$ defined by

$$|\text{map}(xs, ys)|_{MAP} = |xs|.$$

- Define $||\ ||_{MAP}$ by recursion as follows:

$$||\text{map}([\,], ys)||_{MAP} = 0,$$
$$||\text{map}([x|xs], ys)||_{MAP} = f(x) + ||\text{map}(xs, ys)||_{MAP}.$$

Assumption 3 of Theorem 4.9 is satisfied by $||\ ||_{MAP}$. Indeed, consider a ground instance

$$\text{map}([x|xs], [y|ys]) \leftarrow p(x, y), \text{map}(xs, ys).$$

of the recursive clause of program MAP, and observe that:

$||\text{map}([x|xs], [y|ys])||_{MAP} = f(x) + ||\text{map}(xs, [y|ys])||_{MAP} \geq f(x) = |p(x,y)|_P,$
$||\text{map}([x|xs], [y|ys])||_{MAP} = f(x) + ||\text{map}(xs, [y|ys])||_{MAP}$
$\geq ||\text{map}(xs, [y|ys])||_{MAP} = ||\text{map}(xs, ys)||_{MAP}.$

By Theorem 4.9 we conclude that MAP $\cup$ P is recurrent. Moreover, we obtain that a goal $\leftarrow \text{map}(xs, ys)$ is bounded if xs is a list of terms each of which is bounded w.r.t. f. (As expected, a term t is bounded w.r.t f if f is bounded on the set of ground instances of t.) Thus we obtained a modular proof scheme for the parametric program MAP.

Note that there is no relationship between $|\text{map}([x|xs], [y|ys])|_{MAP}$ which equals $|xs| + 1$ and $|p(x,y)|_P$ which equals $f(x)$, so with this natural choice of level mappings we cannot apply here Theorem 4.8.

5 A modular approach to left termination

5.1 Semi-acceptable programs

An analogous modification of the notion of acceptability yields a modular approach to the proofs of left termination.

Definition 5.1 *Let P be a program, $|\ |$ a level mapping for P and I a (not necessarily Herbrand) interpretation of P.*

- *A clause of P is called semi-acceptable w.r.t. $|\ |$ and I, if I is its model and for every ground instance $A \leftarrow \mathbf{A}, B, \mathbf{B}$ of it such that $I \models \mathbf{A}$*

  (i) $|A| > |B|$ if $rel(A) \simeq rel(B)$,
  (ii) $|A| \geq |B|$ if $rel(A) \sqsupset rel(B)$.

- *A program P is called semi-acceptable w.r.t. $|\ |$ and I, if all its clauses are. P is called semi-acceptable if it is semi-acceptable w.r.t. some level mapping and an interpretation of P.*

Again, the use of the premise $I \models \mathbf{A}$ forms the *only* difference between the concepts of semi-recurrence and semi-acceptability.

The following observations are immediate. The first one is a counterpart of Lemma 3.3.

Lemma 5.2 *A program is semi-recurrent w.r.t. $|\ |$ iff it is semi-acceptable w.r.t. $|\ |$ and B_P.*

Lemma 5.3 *If a program is acceptable w.r.t. $|\ |$ and I, then it is semi-acceptable w.r.t. $|\ |$ and I.*

Also, the proof of Lemma 4.4 can be literally viewed as a proof of the following analogous result for semi-acceptable programs.

Lemma 5.4 *If a program is semi-acceptable w.r.t. $|\ |$ and I, then it is acceptable w.r.t. a level mapping $||\ ||$ and the same interpretation I. Moreover, for each atom A, if A is bounded w.r.t. $|\ |$, then A is bounded w.r.t. $||\ ||$.*

The following is a direct consequence of Lemmata 5.3 and 5.4.

Corollary 5.5 *A program is acceptable iff it is semi-acceptable.*

Let us consider now the issue of modularity. The following is an analogue of Theorem 4.6 for semi-acceptable programs.

Theorem 5.6 *Let P and Q be two programs such that no relation occurs in both of them. Suppose that*

- *Q is semi-acceptable w.r.t. level mapping $|\ |_Q$ and interpretation I_Q,*
- *P is semi-acceptable w.r.t. level mapping $|\ |_P$ and interpretation I_P.*

Then $P \cup Q$ is semi-recurrent w.r.t. $|\ |$ and $I_P \cup I_Q$, where $|\ |$ is defined as follows:

$$|A| = \begin{cases} |A|_P & \text{if } rel(A) \text{ is defined in } P, \\ |A|_Q & \text{if } rel(A) \text{ is defined in } Q. \end{cases}$$

Next, note the following analogue of Theorem 4.8 for semi-acceptable programs.

Theorem 5.7 *Let P and Q be two programs such that P extends Q. Suppose that*
(1) *Q is semi-acceptable w.r.t. $|\ |_Q$ and $I_P \cap B_Q$,*
(2) *P is semi-acceptable w.r.t. $|\ |_P$ and I_P,*
(3) *for every ground instance $A \leftarrow \mathbf{A}, B, \mathbf{B}$ of a clause of P such that $I_P \models \mathbf{A}$*
$$|A|_P \geq |B|_Q \text{ if } rel(B) \text{ is defined in } Q.$$

Then $P \cup Q$ is semi-acceptable w.r.t. $|\ |$ and I_P, where $|\ |$ is defined as follows:

$$|A| = \begin{cases} |A|_P & \text{if } rel(A) \text{ is defined in } P, \\ |A|_Q & \text{if } rel(A) \text{ is defined in } Q. \end{cases} \tag{5.1}$$

Proof. The proof is identical to that of Theorem 4.8. ∎

As in the case of semi-recurrent programs we cannot always hope that two unrelated level mappings satisfy condition 3 of this theorem. The following analogue of Theorem 4.9 for semi-acceptable programs deals with this difficulty.

Theorem 5.8 *Let P and Q be two programs such that P extends Q, and let I_P be a model of $P \cup Q$. Suppose that*
(1) *Q is semi-acceptable w.r.t. $|\ |_Q$ and $I_P \cap B_Q$,*
(2) *P is semi-acceptable w.r.t. $|\ |_P$ and I_P,*
(3) *there exists a level mapping $\|\ \|_P$ such that for every ground instance $A \leftarrow \mathbf{A}, B, \mathbf{B}$ of a clause from P such that $I_P \models \mathbf{A}$*
   (a) $\|A\|_P \geq \|B\|_P$ *if $rel(B)$ is defined in P,*
   (b) $\|A\|_P \geq |B|_Q$ *if $rel(B)$ is defined in Q.*

Then $P \cup Q$ is semi-acceptable w.r.t. $|\ |$ and I_P, where $|\ |$ is defined as follows:

$$|A| = \begin{cases} |A|_P + \|A\|_P & \text{if } rel(A) \text{ is defined in } P, \\ |A|_Q & \text{if } rel(A) \text{ is defined in } Q. \end{cases}$$

Proof. The proof is identical to that of Theorem 4.9. ∎

5.2 Examples

We now present some applications of the modular method for proving left termination. In the following we adopt proof outlines also as a proof format for the verification of assumption 3 of Theorems 5.7 and 5.8. We refer to such proof outlines with the qualification *weak*, and assume that for weak proof outlines condition 4 of Section 3.3 is amended as follows, by replacing $>$ by $\geq$:

$4'$. For $i \in [1, n]$: $f_1 \wedge \cdots \wedge f_{i-1} \Rightarrow t_0 \geq t_i$.

5.2.1 Permutation

Reconsider the program PERMUTATION:

```
perm(Xs, Ys) ←
   Ys is a permutation of the list Xs.

perm(Xs, [X | Ys]) ←
   app(X1s, [X | X2s], Xs),
   app(X1s, X2s, Zs),
   perm(Zs, Ys).
perm([], []).
```

augmented by the APPEND program.

Denote the program defining the PERMUTATION relation by PERM. Clearly, PERM extends APPEND, and perm ⊐ app. Recall that APPEND is recurrent w.r.t. $|\text{app}(\text{xs}, \text{ys}, \text{zs})| = \min(|\text{xs}|, |\text{zs}|)$. Observe the following:

- the relation perm is defined by induction on the length of its first argument. Indeed, the program PERM is semi-acceptable w.r.t. $|\ |$ and I_{PERM} defined by:

$$|\text{perm}(\text{xs}, \text{ys})| = |\text{xs}|,$$

$$I_{PERM} = [\text{perm}(\text{Xs}, \text{Ys})]$$
$$\cup \ \{\text{app}(\text{xs}, \text{ys}, \text{zs}) \mid |\text{xs}| + |\text{ys}| = |\text{zs}|\}.$$

The proof that I_{PERM} is a model of APPEND is as in Section 3.3. The following is a proof outline for the semi-acceptability of the recursive clause for perm w.r.t. $|\ |$ and I_{PERM}:

$$\begin{aligned}
&\texttt{perm(xs,[x|ys])} \leftarrow &&\{|\texttt{xs}|\} \\
&\quad \texttt{app(x1s,[x|x2s],xs)}, \\
&\quad \{|\texttt{xs}| = |\texttt{x1s}| + |\texttt{x2s}| + 1\} \\
&\quad \texttt{app(x1s,x2s,zs)}, \\
&\quad \{|\texttt{zs}| = |\texttt{x1s}| + |\texttt{x2s}|\} \\
&\quad \texttt{perm(zs,ys)}. &&\{|\texttt{zs}|\}.
\end{aligned}$$

- Assumption 3 of Theorem 5.7 is satisfied as the following weak proof outline shows:

$$\begin{aligned}
&\texttt{perm(xs,[x|ys])} \leftarrow &&\{|\texttt{xs}|\} \\
&\quad \texttt{app(x1s,[x|x2s],xs)}, &&\{\min(\texttt{x1s,xs})\} \\
&\quad \{|\texttt{xs}| = |\texttt{x1s}| + |\texttt{x2s}| + 1\} \\
&\quad \texttt{app(x1s,x2s,zs)}, &&\{\min(\texttt{x1s,zs})\} \\
&\quad \{|\texttt{zs}| = |\texttt{x1s}| + |\texttt{x2s}|\} \\
&\quad \texttt{perm(zs,ys)}.
\end{aligned}$$

Hence, by Theorem 5.7 and Lemma 5.4 we conclude that PERMUTATION = PERM ∪ APPEND is acceptable w.r.t. | | and I_{PERM}. We thus achieved the same result of Section 3.3, but in a modular way, and using a more natural level mapping for perm.

5.2.2 A divide & conquer scheme

Reconsider the *divide and conquer* schema DC which is parametric w.r.t. the base, conquer, divide and merge relations:

```
dcsolve(X, Y) ←
   base(X),
   conquer(X, Y).
dcsolve(X, Y) ←
   divide(X, X0, X1, X2),
   dcsolve(X1, Y1),
   dcsolve(X2, Y2),
   merge(X0, Y1, Y2, Y).
```

Let P be a program defining the relations base, conquer, divide and merge. Clearly, DC extends P, and dcsolve ⊐ base, conquer, divide, merge. Assume that P is acceptable w.r.t. $|\ |_P$ and I_P defined as follows:

$$|\texttt{base(x)}|_P = ||\texttt{x}||,$$

$$|\text{conquer}(x,y)|_P = ||x||,$$
$$|\text{divide}(x, x0, x1, x2)|_P = ||x||,$$
$$|\text{merge}(x0, y1, y2, y)|_P = ||x0|| + ||y1|| + ||y2||,$$

$$I_P = \quad [\text{base}(X)]$$
$$\cup \quad \{\text{conquer}(x,y) \mid ||x|| \geq ||y||\}$$
$$\cup \quad \{\text{divide}(x, x0, x1, x2) \mid ||x|| \geq ||x0|| + ||x1|| + ||x2||$$
$$\wedge \ ||x|| > ||x1|| \wedge ||x|| > ||x2||\}$$
$$\cup \quad \{\text{merge}(x0, y1, y2, y) \mid ||y|| \leq ||x0|| + ||y1|| + ||y2||\},$$

where $||\ ||$ denotes some function assigning natural numbers to ground terms.

Notice that these assumptions are quite natural for a large class of programs following the divide and conquer paradigm.

We observe the following:

- the program DC is acceptable w.r.t. $|\ |_{DC}$ and I_{DC} defined by:

$$|\text{dcsolve}(x,y)|_{DC} = ||x||,$$
$$I_{DC} = I_P \cup \{\text{dcsolve}(x,y) \mid ||x|| \geq ||y||\}.$$

The proof outline for the non-recursive clause of DC is obvious. For the recursive clause take the following proof outline:

$\{||x|| \geq ||y||\}$
dcsolve(x, y)   ← $\{||x||\}$
   divide(x, x0, x1, x2),
   $\{||x|| \geq ||x0|| + ||x1|| + ||x2||$
   $\wedge \ ||x|| > ||x1|| \wedge ||x|| > ||x2||\}$
   dcsolve(x1, y1), $\{||x1||\}$
   $\{||x1|| \geq ||y1||\}$
   dcsolve(x2, y2), $\{||x2||\}$
   $\{||x2|| \geq ||y2||\}$
   merge(x0, y1, y2, y).
   $\{||x0|| + ||y1|| + ||y2|| \geq ||y||\}.$

- Assumption 3 of Theorem 5.7 is satisfied as the following weak proof outlines show:

dcsolve(x, y)   ← $\{||x||\}$
  base(x), $\{||x||\}$
  conquer(x, y). $\{||x||\}$

dcsolve(x, y) ← $\{||x||\}$
  divide(x, x0, x1, x2), $\{||x||\}$
  $\{||x|| \geq ||x0|| + ||x1|| + ||x2||$
    $\land\ ||x|| > ||x1|| \land ||x|| > ||x2||\}$
  dcsolve(x1, y1),
  $\{||x1|| \geq ||y1||\}$
  dcsolve(x2, y2),
  $\{||x2|| \geq ||y2||\}$
  merge(x0, y1, y2, y). $\{||x0|| + ||y1|| + ||y2||\}$

Using Theorem 5.7 and Lemma 5.4 we conclude that DC∪P is acceptable w.r.t. | | and I_{DC}, where

$$|A| = \begin{cases} |A|_{DC} & \text{if } rel(A) = \text{dcsolve}, \\ |A|_P & \text{otherwise.} \end{cases}$$

Moreover, we obtain that a goal ← dcsolve(x, y) is bounded if dcsolve(x, y) rigid, so in particular if x is ground. Thus we have obtained a modular proof scheme for divide and conquer programs.

As a direct application, note that the program QUICKSORT can be defined as QUICKSORT = DC ∪ P by putting

$$qs \equiv \text{dcsolve},$$

and defining P as follows:

  base([]).
  conquer([], []).
  divide([X|Xs], [X], Littles, Bigs) ←
    part(X, Xs, Littles, Bigs).
  merge([X], Ls, Bs, Ys) ←
    app(Ls, [X|Bs], Ys).

It is easy to check that P satisfies the conditions of the presented proof scheme for DC, and thus we can directly conclude that QUICKSORT is left terminating, and that for a list s, all LD-derivations of QUICKSORT $\cup \{\leftarrow \mathtt{qs(s,t)}\}$ are finite. (To be more precise, we obtain QUICKSORT from the above program by unfolding in the sense of Tamaki and Sato (1984).)

5.2.3 A generate & test scheme

Consider the following one-clause program GT, representing a *generate and test* schema; it is parametric w.r.t. the generate and test relations:

```
gtsolve(X, Y) ←
    generate(X, Y),
    test(Y).
```

Let P be a program defining the relations generate and test. Clearly, GT extends P, and gtsolve $\sqsupset$ generate, test. Assume that P is acceptable w.r.t. $|\ |_P$ and I_P defined as follows, where, as before, $||\ ||$ denotes a function assigning natural numbers to ground terms:

$$|\mathtt{generate(x,y)}|_P = ||\mathtt{x}||,$$
$$|\mathtt{test(y)}|_P = ||\mathtt{y}||,$$

$$I_P|\{\mathtt{test, generate}\} = [\mathtt{test(Y)}]$$
$$\cup \ \{\mathtt{generate(x,y)} \mid ||\mathtt{x}|| \geq ||\mathtt{y}||\}.$$

Here for a Herbrand interpretation I and a set of relations R, we denote by $I|R$ the restriction of I to the relations belonging to R.

We observe the following:

- the program GT is trivially semi-recurrent w.r.t. any level mapping. In fact, the only clause of GT is non-recursive.
- Define $|\ |_{GT}$ and I_{GT} as follows:

$$|\mathtt{gtsolve(x,y)}|_{GT} = ||\mathtt{x}||,$$

$$I_{GT} = I_P \cup [\mathtt{gtsolve(x,y)}].$$

Assumption 3 of Theorem 5.7 is satisfied by $|\ |_P$, $|\ |_{GT}$ and I_{GT}, as the following weak proof outline shows:

$$\begin{array}{rll}
\mathtt{gtsolve(x,y)} & \leftarrow & \{||\mathtt{x}||\} \\
& \mathtt{generate(x,y)}, & \{||\mathtt{x}||\} \\
& \{||\mathtt{x}|| \geq ||\mathtt{y}||\} & \\
& \mathtt{test(y)}. & \{||\mathtt{y}||\}
\end{array}$$

By Theorem 5.7 and Lemma 5.4 we conclude that GT ∪ P is acceptable w.r.t. | | and I_{GT}, where

$$|A| = \begin{cases} |A|_{GT} & \text{if } rel(A) = \text{gtsolve}, \\ |A|_P & \text{otherwise}. \end{cases}$$

Moreover, we obtain that a goal ← gtsolve(x, y) is bounded if gtsolve(x, y) is rigid, so in particular if x is ground. Thus we have obtained a modular proof scheme for generate and test programs.

As a direct application, consider the program SLOWSORT = GT ∪ P, obtained by putting

$$\begin{aligned} \text{ss} &\equiv \text{gtsolve}, \\ \text{generate} &\equiv \text{perm}, \\ \text{test} &\equiv \text{ordered}, \end{aligned}$$

and

$$P = \text{PERMUTATION} \cup \text{ORDERED},$$

where ORDERED is defined by

ordered(Xs) ←
    Xs is a ≤-ordered list of natural numbers.

ordered([]).
ordered([X]).
ordered([X, Y | Xs]) ← X ≤ Y, ordered([Y| Xs]).

ORDERED is clearly recurrent w.r.t. the level mapping |ordered(ys)| = |ys|, so acceptable w.r.t. | | and [ordered(XS)]. By Theorem 5.6 PERMUTATION ∪ ORDERED is acceptable w.r.t. | | defined by

$$\begin{aligned} |\text{perm(xs, ys)}|_P &= |\text{xs}|, \\ |\text{ordered(ys)}|_P &= |\text{ys}|, \end{aligned}$$

and the model [ordered(XS)] ∪ I_{PERM}.

Thus we can directly conclude that SLOWSORT is left terminating and that for a list s, all LD-derivations of SLOWSORT ∪ { ← ss(s, t)} are finite.

5.2.4 *A relational* fold *program*

Consider the following program FOLD which implements a relational equivalent of the higher-order combinator *fold-left* of functional programming. The program FOLD is parametric w.r.t. relation op. We assume that op is the relational equivalent of a binary operator *op*, in the sense that op(x,y,z) holds iff z = (x *op* y).

fold(X, [Y1, ..., Yn], Z) ←
    Z = (... ((X *op* Y1) *op* Y2) ... *op* Yn)

```
fold(X, [Y | Ys], Z) ←
    op(X, Y, V), fold(V, Ys, Z).
fold(X, [ ], X).
```

Let OP be a program defining the relation op, such that FOLD extends OP (hence: fold ⊐ op.) Assume that OP is acceptable w.r.t. | |$_{OP}$ and I_{OP} satisfying the following properties:

$$|\text{op}(x,y,z)|_{OP} = f(x) + g(y)$$
$$I_{OP}|\{\text{op}\} = \{\text{op}(x,y,z) \mid f(x) + g(y) \geq f(z)\},$$

where f, g denote some functions assigning natural numbers to ground terms.

We observe the following:

- the program FOLD is trivially recurrent w.r.t. | |$_{FOLD}$ defined by

$$|\text{fold}(x, ys, z)|_{FOLD} = |ys|.$$

By Lemma 3.3 FOLD is acceptable w.r.t. | |$_{FOLD}$ and $I_{FOLD} = I_{OP} \cup [\text{fold}(x, ys, z)]$.

- Define a function || || assigning natural numbers to ground terms by recursion as follows:

$$||[y|ys]|| = g(y) + ||ys||,$$
$$||x|| = 0 \text{ otherwise.}$$

Assumption 3 of Theorem 5.8 is satisfied by putting

$$||\text{fold}(x, ys, z)||_{FOLD} = f(x) + ||ys||,$$

and using I_{FOLD} defined before. Indeed, the weak proof outline for the non-recursive clause of FOLD is obvious and for the recursive clause we have the following weak proof outline:

fold(x, [y|ys], z) ← $\{f(x) + g(y) + ||ys||\}$
    op(x, y, v), $\{f(x) + g(y)\}$
    $\{f(x) + g(y) \geq f(v)\}$
    fold(v, ys, z). $\{f(v) + ||ys||\}.$

By Theorem 5.8 we conclude that FOLD ∪ OP is acceptable. Moreover, we obtain that a goal

$$\leftarrow \text{fold}(x, ys, z)$$

is bounded if x is bounded w.r.t. f, and ys is a list of terms each of which is bounded w.r.t. g. Thus we have obtained a modular proof scheme for the parametric program FOLD. Notice that with the above choice of the level mappings we cannot apply here Theorem 5.7, since $|\ |_{FOLD}$ is unrelated to $|\ |_{OP}$, whereas $||\ ||_{FOLD}$ does not need to decrease in recursive calls.

As a direct application, consider the program SUMLIST = FOLD $\cup$ OP, obtained by putting OP = SUM, where SUM is defined as in Section 2.3, and

$$op \equiv sum.$$

It is easy to check that OP satisfies the conditions of the presented proof scheme for FOLD, by putting:

$$f(x) = 0,$$
$$g(x) = \text{size}(x).$$

Thus we can directly conclude that SUMLIST is acceptable, and that, for a ground ys, all LD-derivations of a goal $\leftarrow$ fold(x, ys, z) w.r.t. SUMLIST are finite. Note that the goal $\leftarrow$ fold(0, ys, z) computes the sum of the elements of the list ys.

5.2.5 *The* MAP *program revisited*

Reconsider the program MAP:

```
map([X1, ..., Xn], [Y1, ..., Yn]) ←
    p(Xi, Yi) holds for i ∈ [1,n].
map([X|Xs], [Y|Ys]) ←
    p(X, Y), map(Xs, Ys).
map([], []).
```

Relax the assumptions made in Section 4.4 on P by assuming that P is acceptable w.r.t. $|\ |_P$ defined as in Section 4.4, and *any* model I of P. It is immediate to observe that the proof outlines of Section 4.4 remain valid with the new assumptions. Hence, by Theorem 5.8, we conclude that MAP $\cup$ P is acceptable; moreover, we obtain the same class of bounded goals as in Section 4.4. Again, we cannot apply here Theorem 5.7, since the level mappings for map and p are unrelated.

5.2.6 *A map coloring program*

Finally, consider a jewel of PROLOG – the following MAP_COLOR program from Sterling and Shapiro (1986, page 212) which generates a coloring of a map in such a way that no two neighbors have the same color. Below we call such a coloring *correct*. The map is represented as a list of regions and colors as a list of available colors. In turn, each region is determined by its name, color and the colors of its neighbors, so it is represented as a term region(name, color, neighbors), where neighbors is a list of colors of the neighboring regions.

```
color_map(Map, Colors) ←
```

Map is correctly colored using Colors.

color_map([Region | Regions], Colors) ←
   color_region(Region, Colors),
   color_map(Regions, Colors).
color_map([], Colors).

color_region(Region, Colors) ←
   Region and its neighbors are correctly colored using Colors.

color_region(region(Name, Color, Neighbors), Colors) ←
   select(Color, Colors, Colors1),
   subset(Neighbors, Colors1).

augmented by the SELECT program.

augmented by the SUBSET program.

Denote by CM the program consisting of the two clauses defining the relation color_map, and by CR the program consisting of the clause defining the relation color_region. Clearly, CM extends CR, and CR extends SELECT and SUBSET. Moreover, color_map ⊐ color_region ⊐ select, subset in the program MAP_COLOR = CM ∪ CR ∪ SELECT ∪ SUBSET.

First we deal with the program CR ∪ SELECT ∪ SUBSET. To this end Theorems 5.6 and 5.7 will be of help. Recall that SELECT is recurrent w.r.t. $|\mathrm{select}(x, xs, ys)| = |xs|$, and that SUBSET is recurrent w.r.t. $|\mathrm{subset}(xs, ys)| = |xs| + |ys|$ and $|\mathrm{member}(x, xs)| = |xs|$. Observe the following:

- the program CR is trivially semi-recurrent w.r.t. any level mapping.
- The Herbrand interpretation $I_S = \{\mathrm{select}(x, xs, ys) \mid |xs| \geq |ys|\}$ is a model of SELECT, as the following proof outlines show:

$$\{1 + |xs| \geq |xs|\}$$
$$\mathrm{select}(x, [x|xs], xs).$$

$$\{1 + |xs| \geq 1 + |ys|\}$$
$$\mathrm{select}(x, [y|xs], [y|ys]) \quad \leftarrow$$
$$\mathrm{select}(x, xs, ys).$$
$$\{|xs| \geq |ys|\}.$$

Consequently, by Lemma 3.3 and Theorem 5.6 SELECT ∪ SUBSET is semi-acceptable w.r.t. | | and $I_S \cup [\mathrm{subset}(\mathrm{Xs}, \mathrm{Ys})] \cup [\mathrm{member}(\mathrm{X}, \mathrm{Xs})]$.

- The programs CR and SELECT ∪ SUBSET satisfy assumption 3 of Theorem 5.7 by putting

$$I_{CR} = I_S \cup [\text{color\_region}(R, Cs)] \cup [\text{subset}(Xs, Ys)] \cup [\text{member}(X, Xs)]$$

and extending | | as follows:

$$|\text{color\_region}(\text{region}(n, c, ns), cs)| = |ns| + |cs|,$$
$$|\text{color\_region}(x, cs)| = 0 \text{ if } x \neq \text{region}(n, c, ns).$$

The associated weak proof outline follows:

$$\text{color\_region}(\text{region}(n, c, ns), cs) \leftarrow \{|ns| + |cs|\}$$
$$\text{select}(c, cs, c1s), \qquad \{|cs|\}$$
$$\{|cs| \geq |c1s|\}$$
$$\text{subset}(ns, c1s).\{|ns| + |c1s|\}.$$

Therefore, by Theorem 5.7, the program CR ∪ SELECT ∪ SUBSET is semi-acceptable w.r.t. | | and I_{CR}.

Now we can deal with the program MAP_COLOR. For this purpose Theorem 5.8 will be of use. Observe the following:

- The program CM is trivially recurrent w.r.t. $|\text{color\_map}(rs, cs)| = |rs|$.
- Define a function || || from lists of regions to natural numbers by induction as follows:

$$||[\text{region}(n, c, ns)|rs]|| = |ns| + ||rs||,$$
$$||[x|rs]|| = ||rs|| \text{ if } x \neq \text{region}(n, c, ns),$$
$$||x|| = 0, \text{ otherwise.}$$

The programs CM and CR ∪ SELECT ∪ SUBSET satisfy assumption 3 of Theorem 5.8 by putting

$$I_{CM} = I_{CR} \cup [\text{color\_map}(Rs, Cs)],$$
$$||\text{color\_map}(rs, cs)||_{CM} = ||rs|| + |cs|.$$

Two weak proof outlines covering all ground instances of the recursive clause of color_map follow. We assume that $x \neq \text{region}(n, c, ns)$.

$$\text{color\_map}([\text{region}(n, c, ns)|rs], cs) \leftarrow \{|ns| + ||rs|| + |cs|\}$$
$$\text{color\_region}(\text{region}(n, c, ns), cs), \qquad \{|ns| + |cs|\}$$
$$\text{color\_map}(rs, cs). \qquad \{||rs|| + |cs|\}$$

$$\text{color\_map}([x|rs], cs) \leftarrow \{||rs|| + |cs|\}$$
$$\text{color\_region}(x, cs),$$
$$\text{color\_map}(rs, cs). \qquad \{||rs|| + |cs|\}.$$

Consequently, by Theorem 5.8, we conclude that the program MAP_COLOR = CM ∪ CR ∪ SELECT ∪ SUBSET is semi-acceptable. Moreover, we obtain that a goal ← color_map(rs, cs) is bounded if cs is a list and rs is a list of regions [region(n_1, c_1, ns_1), ..., region(n_k, c_k, ns_k)], where each ns_i ($i \in [1, k]$) is a list. Thus, MAP_COLOR terminates for the desired class of goals.

Acknowledgements

This research was partly supported by the ESPRIT Basic Research Action 6810 (Compulog 2).
We thank Antonio Brogi, Augusto Ciuffoletti and Paolo Mancarella for useful discussions on the final version and Andrea Schaerf for helpful comments.

Bibliography

1. K. R. Apt (1990). Logic programming. In J. van Leeuwen, editor, *Handbook of Theoretical Computer Science*, pages 493–574. Elsevier, Amsterdam. Vol. B.
2. K. R. Apt and D. Pedreschi (1990). Studies in pure Prolog: termination. In J.W. Lloyd, editor, *Symposium on Computational Logic*, pages 150–176. Springer-Verlag, Berlin.
3. K. R. Apt and D. Pedreschi (1993). Reasoning about termination of pure PROLOG programs. *Information and Computation*, 106(1): 109–157.
4. M. Bezem (1989). Characterizing termination of logic programs with level mappings. In E. L. Lusk and R. A. Overbeek, editors, *Proceedings of the North American Conference on Logic Programming*, pages 69–80. MIT Press, Cambridge, MA.
5. M.A. Bezem (1993). Strong termination of logic programs. *Journal of Logic Programming*, 15(1 & 2):79–98.
6. L. Cavedon (1989). Continuity, consistency, and completeness properties for logic programs. In G. Levi and M. Martelli, editors, *Proceedings of the Sixth International Conference on Logic Programming*, pages 571–584. MIT Press, Cambridge, MA.
7. N. Dershowitz (1987). Termination of rewriting. *Journal of Symbolic Computation*, 8:69–116.
8. D. De Schreye, K. Verschaetse, and M. Bruynooghe (1992). A framework for analyzing the termination of definite logic programs with respect to call patterns. In *Proceedings of the International Conference on Fifth*

Generation Computer Systems 1992, pages 481–488. Institute for New Generation Computer Technology (ICOT), Tokyo, Japan.

9. D. König (1927). Über eine Schlußweise aus dem Endlichen ins Unendliche. *Acta Litt. Ac. Sci.*, 3:121–130.

10. J. W. Lloyd (1987). *Foundations of Logic Programming*. Springer-Verlag, Berlin, second edition.

11. U.S. Reddy (1986). On the relationship between logic and functional languages. In D. DeGroot and G. Lindstrom, editors, *Functional and Logic Programming*, pages 3–36. Prentice-Hall, Englewood Cliffs, NJ.

12. L. Sterling and E. Shapiro (1986). *The Art of PROLOG*. MIT Press, Cambridge, MA.

13. H. Tamaki and T. Sato (1984). Unfold/Fold Transformations of Logic Programs. In Sten-Åke Tärnlund, editor, *Proc. Second International Conference on Logic Programming*, pages 127–139.

14. J. D. Ullman and A. van Gelder (1988). Efficient tests for top-down termination of logical rules. *J. ACM*, 35(2):345–373.

5
Logic + Control Revisited: an Abstract Interpreter for GÖDEL Programs

Egon Börger
Università di Pisa

Elvinia Riccobene
Università di Catania

Abstract

We develop a simple interpreter for programs of the new logic programming language GÖDEL. The definition provides a clean interface between logical and control components for execution of GÖDEL programs. The construction is given in abstract terms which cover the general logic programming paradigm and allow for concurrency.

The formalization directly reflects the intuitive procedural understanding of programs, but is formulated at the level of abstract search spaces and proceeds in a modular fashion. This combination of procedural and abstract features, made possible by use of Gurevich's notion of *evolving algebras*, provides a tool for mathematical (machine and proof system independent) description and analysis of design decisions for logic programming languages; it also lays the ground for provably correct stepwise refinements, through a hierarchy of specifications at lower levels, down to implementations.

1 Introduction

Hill and Lloyd (1992) have proposed the new general-purpose logic programming language GÖDEL, with particular emphasis on improving the declarative semantics compared with PROLOG. GÖDEL has a type system which is based on many-sorted logic with parametric polymorphism, a module system and infinite precision integers, rationals and floating-point numbers; it can solve constraints over finite domains of integers as well as linear rational constraints; it supports processing of finite sets as well as metalogical facilities (created in order to provide support for metaprograms for analysis, transformation, compilation, verification and debugging of programs).

We do not analyze any of these features but concentrate our attention in this paper on the special role which is played by three forms of nondeterminism in the logic programming paradigm (see below) that characterize the top level definition of the procedural semantics of GÖDEL programs. From the implementation point of view, nondeterminism reflects the intended flexibility of GÖDEL's computational rule and the desire "to give implementors the option" of not relying upon (some generalization of) SLDNF-resolution but "of using other theorem proving techniques to implement the language, e.g. ones which avoid floundering or are more complete" (Lloyd 1992).

We develop here, by stepwise refinement, a mathematically precise but simple procedural formalization of the language which describes the full (control flow) behavior of GÖDEL programs on the basis of *abstract* (machine and resolution independent) *search spaces*. In particular our specification provides a rigorous basis for an equivalence proof between declarative and procedural semantics of pure Gödel programs. We exemplify the model for (an SLDNF like) resolution as basic computation mechanism; we do it in a modular way and exhibit explicitly the interface where one can

- adapt the semantics to possible future changes in the design of the language,
- refine it in terms of particular proof systems or lower level specifications which are driven by consideration of execution efficiency and similar implementation issues.

Our search spaces are reminiscent of *PROLOG tree algebras*, which appear in hybrid stack oriented form in (Börger 1990b) and have been defined in (Börger and Rosenzweig 1991a) and used in (Börger and Rosenzweig 1994a) as basis for a formal model of PROLOG. The abstract search spaces have been obtained from PROLOG tree algebras by a further abstraction step, introduced in order to directly reflect the desired flexible character of GÖDEL's computational rule.

This rule is indeed nondeterministic

- in choosing where the next deduction step takes place, thus abstracting in particular from PROLOG's depth-first strategy,
- in selecting, out of a conjunction of goals, the literal to be computed, thus abstracting from PROLOG's left-to-right strategy ,
- in selecting a clause to reduce the current call, thus abstracting from PROLOG's sequential strategy and from any scheme for indexing, switching, last call optimization etc.

Gurevich's concept of *external* functions (Gurevich 1991) in *evolving algebras* (Gurevich 1988) gave us the technical instrument to express in an explicit and transparent way the role of these three (pairwise orthogonal) nondeterministic control components for the semantics of GÖDEL (and in general logic) programs. The effect of control by delaying, pruning, conditionals, negation, etc. on program execution is expressed by abstract conditions, which can be viewed as directives for the implementation of efficient but semantically correct search strategies. In this way a highly abstract mathematical framework for logic programming systems is laid down in which one can conduct a comparative study of the different answers which have been given, by well known logic programming languages and their implementations, to the *Gretchenfrage* of logic *programming*, namely to explicitly and exactly determine the real relation between "logic" and "control", between declarative (high-level) specification and procedural ("low"-level) implementation of (not only pure) logic programs.

On this basis we give a simple description of the high-level procedural semantics of GÖDEL programs, including, besides a full treatment of user-defined predicates, as characteristic built-in control flow features the *pruning* operator (which relativizes PROLOG's *cut* operator and generalizes the *commit* operator of the concurrent logic programming languages), *negation* and *conditionals* together with their *delay* properties. Other system features in GÖDEL (for constraints, number manipulation etc.) can be dealt with in this model by adapting the methodology developed in (Börger 1990b, 1992; Börger and Rosenzweig 1994a; Börger and Schmitt 1991) for the formalization of their (sometimes quite different) PROLOG homonyms; see also (Beierle and Börger 1992) for a treatment of polymorphic types for an extension of PROLOG. We indicate the effect of depth-first search on this abstract specification and briefly discuss how to obtain from it a concurrent model of GÖDEL.

2 Evolving algebras

We expect from the reader basic knowledge of logic programming (Apt 1990; Lloyd 1987) or GÖDEL (Hill and Lloyd 1992) and rudimentary knowledge of the language of first order logic.

Our model comes in the form of *evolving algebras*. This concept has been introduced by Gurevich (1988) and has since then been applied successfully for the specification of languages covering all the major programming paradigms; see (Gurevich and Morris 1988; Börger 1990a,b,1992; Gurevich and Moss 1990; Börger and Schmitt 1991; Gurevich and Huggins 1993; Börger and Rosenzweig 1994a; Beierle and Börger 1992; Börger and Riccobene 1993; Börger and Riccobene 1992; Börger et al. 1994; Blakley 1992; Gottlob et al. 1991). Although evolving algebras have a rigorous mathematical foundation as transition systems over first order structures (see Gurevich 1991; Glavan and Rosenzweig 1993), we make this paper independent by listing in this section some definitions which will enable the reader to grasp our rules as "pseudocode over abstract data".

For a general and extensive discussion of the evolving algebra methodology for specification of (the semantics of) programming languages and systems, in particular in comparison to other well-known approaches in the literature (denotational, algebraic, axiomatic, ...), we refer the reader to the introduction to (Börger and Rosenzweig 1994a).

The abstract data come as elements of (not further analyzed) sets (domains, *universes*). The operations allowed on universes will be represented by partial *functions*. The setup is allowed to *evolve* in time, by executing *function updates* of form

$$f(t_1, \ldots, t_n) := t$$

whose execution is to be understood as *changing* (or defining, if there was none) the value of function f at given arguments.

The 0-ary functions are like variable quantities of ancient mathematics or *variables* of programming; this is the reason why we abstain from calling them *constants*. Functions which do not appear as outer function f on the left-hand side of function updates have been called *external* functions in (Gurevich 1991). They are not changed by the rule system, but they might nevertheless be (supposed to be) subject to change due to the activity of some external agent; in this case they are called *dynamic* external functions. In this paper we will make crucial use of such dynamic external functions.

For a natural high-level formalization of Gödel's pruning operator we allow also parameterized and guarded function updates of form

$$U(\vec{j}) \text{ or } \forall j(guard(\vec{j})) : U(\vec{j})$$

where U is a function update in which parameters $\vec{j}$ may occur, meaning that $U(\vec{j})$ is executed for all values of parameters (which satisfy $guard(\vec{j})$). (Typically j will be used to range over nodes on a segment of a path in a tree.) Note that such updates are not considered in the original definition given by (Gurevich 1988).

We shall also allow some of the universes (typically initially empty) to *grow* in time, by executing updates of form

Extend U **by** $t_1, \ldots, t_n$ **with** *updates* **EndExtend**

where *updates* may (and should) depend on the t_i's, setting the values of some functions on the *newly created* elements t_i of U.

The precise way our "abstract machine" (evolving algebras) may evolve in time will be determined by a finite set of *transition rules* of form

If *condition* **then** *updates*

where *condition* is a (usually Boolean) expression (guard), the truth of which triggers *simultaneous* execution of all updates listed in *updates*. Simultaneous execution helps us to avoid coding, for example to interchange two values. Note that since functions may be partial, equality in the guards is to be interpreted in the usual sense of "partial algebras", as implying that both arguments are defined; for more details see (Gurevich 1991; Glavan and Rosenzweig 1993).

In applications an evolving algebra usually comes together with a set of *integrity constraints*, i.e. extralogical axioms and/or rules of inference, specifying the intended domains. Indeed we are usually interested only in states which are reachable from some designated *initial states*, specified using any formal methods, such as those of algebraic specification.

The forms obviously reducible to the above basic syntax, which we shall freely use as abbreviations, are **let** and **if then else**. We shall assume that we have the standard mathematical universes of Booleans, integers, lists of whatever, etc. (as well as the standard operations on them) at our disposal without further mention.

For the sake of exposition the model we are going to construct specifies the behavior of GÖDEL programs and goals such that the module structure is flattened and, when all commits are removed from the body of each statement, what remains is a normal program and a normal goal (which may contain the conditional construct). The additional first-order features, other connectives and quantifiers, which are allowed in Gödel statements, can easily be described as built-in predicates (using the technique presented here for negation and conditionals) or by preprocessing (which reduces to normal clause form, see (Lloyd 1987), and are skipped here for space reasons.

For simplicity we assume programs and queries to be well (static) type-checked[17].

3 Signature of GÖDEL algebras

This and the next section explain the basic data types (domains and functions) which are used in the transition rules of Section 5.

A GÖDEL computation can be seen as systematic, weakly controlled nondeterministic search of all solutions to an initially given query, to be found in a space of possible solutions. For a formalization we represent the set of computation states by a set of nodes, a subset of which will be structured as a tree to reflect the control conditions (like pruning) imposed by GÖDEL on angelic nondeterminism, an idea used already in (Börger 1990b) (in hybrid stack oriented form) and in (Börger and Rosenzweig 1991a, 1994a) as basis for defining backtracking in an abstract model for PROLOG. Therefore *GÖDEL algebras* have as basic universe a set **Node** representing all possible *GÖDEL computation states*, containing a distinguished element (0-ary function) *currnode* representing the current computation state.

One aspect of GÖDEL's nondeterminism is that the computation can leave what it is doing at any part of the search space and compute at another part, maybe returning later; this abstracts from any particular strategy, like depth-first or breadth-first or others, in building and searching

---

[17]See (Beierle and Börger 1992) for a formal specification, using the methodology of evolving algebras, of polymorphic types present at runtime for an extension of PROLOG.

the solution tree, and brings GÖDEL computations rather close to the logician's understanding of deductions. We reflect this nondeterminism explicitly by treating *currnode* as dynamic external function. This means that *currnode* will never be updated by any of our rules, but is nevertheless supposed to change value due to some outside (hidden or implementation defined) control. The reader might be helped by considering *currnode* as a demon who executes a basic logical computation step. The control determines not the logic of what the demon does, but the place where he is put to work. This control reflects some restrictions to full nondeterminism (such as that no step should be repeated, that pruning should be respected etc.), which we will formulate explicitly below [18].

Each element n of **Node** has to carry the relevant information for a complete description (at the abstraction level of the user of the language which interests us here) of the computation state it represents, thus allowing description of the execution of its subsequent computation in the subtree rooted at n. At the desired abstraction level this information consists of the literals which have still to be computed, the substitution computed (or the constraint system accumulated) so far and the statements which are still to be considered as candidates for alternative computations. We formalize these concepts by three functions

$$goal : \textbf{Node} \to \textbf{Goal}$$

(associating with each node the goal which is still to be computed)

$$sub : \textbf{Node} \to \textbf{Subst}$$

(representing the substitution (or the sytem of constraints) current at a state)

$$cands : \textbf{Node} \to \textbf{Clause}^*$$

(representing the relevant candidate clauses at given state), where **Clause*** denotes (a subset of) the power set of the set **Clause** of (user-defined) program statements.

Goal is the set of (normal) GÖDEL goals, i.e. terms constructed applying the (*and*) operator "&" to the empty goal and elements of universes **Lit, Conditional** and **Commit**. **Lit** denotes the universe of (positive and negative) literals; **Conditional** denotes the set of GÖDEL *conditionals*

---

[18]Thus we decide to exclude, from the specification of the semantics of GÖDEL programs at this level of abstraction, any explication of the dynamics of *currnode*. This means in particular that we can avoid introducing the tree structure on **Node** into the signature of GÖDEL algebras, say by a partial function *parent* : **Node** → **Node** (undefined on *root*), to be dynamically updated through rules in such a way that from each node (different from *root*) there is a unique *parent* path towards *root*. We can define instead at the metalevel, in terms of applications of our rules, a parent-relation which will allow us to speak of **Node** as a tree. This will be useful for the description of pruning.

and comes with three functions *cond, then, else* which on arguments IF *Cond* THEN *Formula1* ELSE *Formula2* yield the conditional's guard, THEN part, ELSE part respectively, where *Cond, Formulai* belong to the universe **Goal**. **Commit** is the universe of terms involving the *commit* operator, i.e. terms of the form $\{g_1\&\cdots\&g_n\}\_l$, $n \geq 0$, where l is (an integer defining) the *label* of the commit and $g_i \in$ **Lit** ∪ **Conditional** ∪ **Commit**. Note that commits are not allowed to occur inside conditionals. **Term** is the set of (generic) GÖDEL terms which contains **Lit, Conditional, Commit, Goal** as subsets.

Subst is a set of (not further specified) *substitutions* and comes together with three abstract functions

$$unify : \textbf{Term} \times \textbf{Term} \rightarrow \textbf{Subst} \cup \{nil\}$$

associating with two terms either their unifying substitution or the answer that there is none; the substitution applying function

$$subres : \textbf{Term} \times \textbf{Subst} \rightarrow \textbf{Term}$$

yielding the result of applying the given substitution to a given term; substitution concatenation

$$\circ : \textbf{Subst} \times \textbf{Subst} \rightarrow \textbf{Subst}.$$

Note that the presentation in terms of substitutions does not limit the generality of our definitions. The latter apply indeed to arbitrary constraint systems by replacing unifiability of terms by solvability of constraints, unification by constraint solving and concatenation of substitutions by accumulation of constraints. (See (Börger and Schmitt 1991; Beierle and Börger 1992) for a detailed account of abstract constraint handling in the framework of evolving algebras.)

Clause comes with auxiliary functions *clhead* : **Clause** → **Lit**, *clbody* : **Clause** → **Goal** yielding *head* and *body* of GÖDEL statements. The current program is represented by a distinguished element *db* (database) of a universe **Program**. The candidate clauses which have to be considered in a call to define the alternatives of a computation state, are accessed using an abstract function

$$procdef : \textbf{Lit} \times \textbf{Program} \rightarrow \textbf{Clause}^*$$

which yields the (normal) statements defining, in the given program, the predicate having the same functor as the given literal.

For reasons of transparency we separate two different aspects of the action to be taken at *currnode* in executing a call: to collect the candidate statements for the potential alternative computation states and to select one of them for execution, distinguished by values *call, select* of a monadic function *mode* with domain **Node**[19]. To be able to speak about *termination* we will use distinguished elements *root* (with the obvious intended meaning) and *stop* with values in {0,1,-1} to indicate running of the system, stop with success and stop due to final failure respectively.

4 Delaying computations

The GÖDEL computation rule[20] is partly built into the system and partly under the control of the programmer through explicit DELAY control declarations. Let **DelayDecl** be the set of DELAY declarations and *delaydecl(db)* the set of delay declarations occurring in program *db*. To describe GÖDEL's goal selection abstractly (thus leaving it largely implementation dependent) let

$$gl\_select : \mathbf{Goal} \times \mathbf{DelayDecl} \times \mathbf{Subst} \to \mathbf{Lit} \cup \mathbf{Conditional}$$

be a (partial) function which selects from a given goal either a positive literal—which under the given substitution is non-delayed according to the given DELAY declarations—or a negative literal (which is ground) or a conditional (with closed guard formula)[21]. (Note that the selected literal or conditional may occur within the scope of a *commit*.) If in the current goal, wrt the current substitution and the current program, there is no such "non-delayed" literal or conditional, the computation *flounders*. *gl_select* reflects this abstractly by not being defined in this case, thus preventing applicability of any transition rule of our system because the value of *act* in the rule guards is undefined, where *act* (for *activator*) stands for

$$act \equiv gl\_select(goal(currnode), delaydecl(db), sub(currnode)).$$

The transition rules in Section 5 can be read in terms of *act*, abstracting from the details of delay specification. Thus our rules apply to any goal selection mechanism of any logic programming system. A natural refinement step for GÖDEL programs consists in further specifying the notion of delayed positive literal, used in the above abstract specification of *gl_select* (and therefore of *act*). Following (Hill and Lloyd 1992) the DELAY declaration *delaydecl(l,P)* for a predicate (literal) *l* in a program *P* has form DELAY $Atom\_1$ UNTIL $Cond\_1$ ... $Atom\_n$ UNTIL $Cond\_n$ with conditions built up

---

[19] For an abstract formulation of the restrictions imposed upon angelic nondeterminism by pruning, negation and conditionals, it will be useful to introduce an additional mode value *abandoned*, see below.

[20] If one wants to include at this level also the constraint solving part, this could abstractly be dealt with incorporating the method of (Börger and Schmitt 1991).

[21] The present version of GÖDEL really does *not specify any more that negative literals must be ground or that the condition in a conditional must be closed. The implementation just has to implement correctly the declarative meaning. However, these are the likely restrictions implementations will make in the near future.* Cited from (Lloyd 1993)

from predicates NONVAR, GROUND, TRUE using conjunction and disjunction, and atoms pairwise without common instance. A set d of DELAY declarations delays calls as follows:

- an atom l is *delayed*, if it has a common instance with some $Atom\_i$ in a DELAY declaration in d without being an instance of $Atom\_i$;
- an atom l is *delayed*, if it is an instance of an $Atom\_i$ in a DELAY declaration in d whose corresponding condition $Cond\_i$ is not satisfied under the instantiating substitution;

It is a routine task to formalize this definition for a refinement of gl_select.

5 Transition rules

We now define the rules by which the system, starting from an algebra with $stop = 0$ (we tacitly assume this running condition to be part of each rule guard) tries to reach successful execution of the query signalled by say $stop = 1$. To abbreviate, we usually suppress the parameter $currnode$ simply by writing

$$\begin{aligned} mode &\equiv mode(currnode) \\ cands &\equiv cands(currnode) \\ s &\equiv sub(currnode) \\ goal &\equiv goal(currnode). \end{aligned}$$

5.1 Initialization and rules for user-defined predicates

We do not assume GÖDEL algebras to be given as static (infinite) search space; they rather evolve dynamically along the computation, starting from initial algebras determined by given program P and query Q. Hence the *initial (ization of) GÖDEL algebras*: **Node** = $\{root, currnode\}$, $goal := $ Q, $mode := call$, $s := \emptyset$, $db := $ P, $stop := 0$. $cands$ is not (yet) defined at $currnode$.

The basic computation step, applicable to *user-defined* predicates, is split into *calling* the activator (to look for the candidate statements for alternative computations at $currnode$) and *selecting* one of them for execution. We will correspondingly have two rules: a *call rule* and a *selection rule*.

The following *call rule*, applicable to nodes with user defined activator in *call* mode, will store the relevant candidate statements (e.g. clauses whose head may unify with act, or whose constraints are consistent with the set of already accumulated constraints), copying from the procedure definition of act into $cands(currnode)$. To ensure that this action of candidate collection takes place at each n at most once, when n is first visited (becomes value of $currnode$ in $call$ mode), the mode turns to $select$.

If $mode = call$ & $is\_user\_defined(act)$
then $cands := procdef(act, db)$

$$mode := select$$

The following *selection rule*, applicable to nodes with user defined activator and in *select* mode, attempts to execute a candidate statement, to be found in *cands(currnode)* (the set of remaining alternatives), using an abstract selection function[22]

$$alt\_select : \textbf{Clause}^* \to \textbf{Clause}.$$

Since this selection function too is kept abstract (considered as implementation defined), the formalization given here for GÖDEL applies *mutatis mutandis* to any logic programming system.

If a candidate statement has been selected, it is erased from the list of candidates (and therefore will not be tried again). If the selected statement does not apply successfully to *act* (typically, in the case of resolution, because the renamed head of the selected clause does not unify with the activator), nothing else is done. Otherwise the selected statement is applied: a new node in *call* mode is created [23], its goal is defined (in the case of resolution by executing the resolution step, replacing the activator by clause body and applying the unifying substitution) and its substitution is updated (in the resolution case by the unifier between the clause head and the calling literal). Note that *currnode* remains in select mode, so that at this node further candidate children may be created.

If $mode = select$ & $is\_user\_defined(act)$
then let $alt \equiv alt\_select(cands)$
  $cands := cands \setminus \{alt\}$
  If $apply(alt, act) \neq nil$
  then Extend Node by t **with**
    $mode(t) := call$
    define $goal(t)$ using $apply(alt, act)$
    update sub at t
  EndExtend

---

[22] For PROLOG it is well known that due to sequential execution of program clauses, different (even consecutive) occurrences of the same clause in a program may have different effect during a computation. For this reason a faithful description of PROLOG has to distinguish between occurrences of clauses and the clauses themselves, see (Börger 1990a; Börger and Rosenzweig 1994a). The introduction of an abstract clause selection function allows hiding of this distinction, at this level of abstraction, and reduction of the question of different occurrences of a same clause to whether *alt_select* and thereby *procdef* have as range a set or a multiset.

[23] Each time we introduce a new node, if not otherwise stated we will consider the present value of *currnode* as its parent node, thus providing the above mentioned tree structure on **Node** which will be needed to deal with GÖDEL's pruning operator. To reflect the desired nondeterminism of GÖDEL in an unspoiled way, we do *not* define this tree structure by updates $parent(t) := currnode$ in the rules.

The function *apply(alt,act)* and the updates *define goal(t) using θ*, *update sub at t* are deliberately kept abstract here; they depend on the basic computation mechanism. In case of resolution they can be specified further by:

$$apply(alt,act) \equiv unify(act,Hd)$$
$$define\ goal(t)\ using\ \theta \equiv goal(t) := (goal[act \leftarrow Bdy])\ \theta$$
$$update\ sub\ at\ t \equiv sub(t) := s \circ \theta$$

where
$$Hd \leftarrow Bdy \equiv rename(alt, currnode)$$

$$\theta \equiv unify(act, Hd)$$

$G[lit \leftarrow term]$ is the result of replacing (the occurence of) *lit* in goal G by *term*, and
$$rename : \mathbf{Term} \times \mathbf{Node} \to \mathbf{Term}$$
associates with a term a new copy of it where all variables *and all commit labels* are renamed at the level determined by the given node. (Thus we handle variable renaming abstractly, avoiding details of term and variable representation.)

Obviously one has to reflect the correctness assumption for applications of Hilbert's ε-operator, namely that the selection function is not applied to an empty domain (here *cands*). One possibility is to let the system check this condition in the form of a guard *If cands = [ ]*, whose satisfaction in PROLOG triggers backtracking (Börger 1990b). If one wishes to keep such control features outside the system (as belonging to implementation and not to semantics), one can impose the correct use of the selection mechanism by an external condition on how *currnode* is allowed to be chosen. Such a condition can be viewed as a directive to be respected by an implementor. The following rather weak condition suffices here:

ε-choice condition: *currnode* never assumes as value a node in mode *select* with empty *cands*.

This condition avoids also repetitions (i.e. trying to apply statements in a part of the deduction which has already been explored.)

The *query success rule* stops the computation with successful halt when hitting the empty goal □: **If** *mode = call* & *goal = □* **then** *stop:= 1* [24].

5.2 The pruning operator

In this section we refine the system to include GÖDEL's pruning operator, called *commit*, of form *{Formula}_l* (with bar commit | of concurrent

---

[24] This rule only stops the system, without giving output or looking for further solutions. These features can easily be provided using the substitution associated with the halting state.

programming languages and one solution commit $\{\ldots\}$ as special case). The brackets $\{\ldots\}$ indicate the *scope* of the commit inside a statement, the label l the scope of the commit over the statements in a procedure definition. The computational effect of an l-commit $\{Formula\}\_l$ is that once *a* solution is found for *Formula*, (a) each potential alternative solution for *Formula* is skipped, and (b) all potential alternatives are skipped which arise from other statements (in the underlying procedure definition) containing a commit with the same label l[25].

To perform actions (a) and (b) one has to keep track of the point in the computation where label l has been introduced into the current goal: the value of *currnode* when, in a selection rule application, l enters the newly updated goal as (renaming of a) label occurring in the body of the selected statement. Since this node will be the starting point of the pruning operation realizing (a) and (b), to be executed on the computation tree once the formula in the scope of an l-commit has been satisfied, we call it *pruning point* and denote it by a function

$$prun\_pt : \textbf{Integer} \rightarrow \textbf{Node}$$

whose values are set by updates

$$prun\_pt(l) := currnode,$$

to be added to the selection rule for each l occurring as (renamed and therefore fresh) commit label in the body of the selected input statement[26].

In order to satisfy stipulation (a), when a commit $\{Formula\}\_l$ has become empty at *currnode* (through successful computation of the formula in the scope), one has to inspect the path[27] from the introduction of the commit—namely at $prun\_pt(l)$— to *currnode* and to abandon each[28] branch which has been or still may be created for an alternative solution of *Formula*. Such branches are created for a potential alternative solution of an activator which occurred during the (successful) computation of *Formula*. Consider the potential children at each node n on the path which has its activator in the scope of a commit with label l. Those children which either will be or have been created and are not on the path must never be

---

[25] From the declarative viewpoint the programmer might be asked to use pruning only where semantically correct, i.e. to prevent parts of the tree which are known to compute no new solutions from being explored. Any statement which relates "declarative" to "procedural" program meaning has to take into account the possibly semantical effect of pruning.

[26] This is an adaptation of the *goal decoration by cutpoints* introduced in (Börger 1990a). For the initialization it seems reasonable to assume $prun\_pt(l) = root$ for each commit label l in the initial query, and to define *root* as parent of the initial value of *currnode*.

[27] It is here that the tree structure of **Node**, defined at the metalevel as indicated above, is used.

[28] We describe eager pruning.

selected. Thus we abandon those children that exist and prevent the remaining clauses in *cands(n)* from creating new children. Using the wording of (Hill and Lloyd 1992) we access these nodes and candidate clauses, which may have to be abandoned and deleted respectively, by abstract functions

$$children\_1st\_kind : \textbf{Node} \times \textbf{Integer} \rightarrow \textbf{Node}^* \cup \{nil\}$$

and

$$cands\_1st\_kind : \textbf{Node} \times \textbf{Integer} \rightarrow \textbf{Clause}^*$$

(It would be easy to define these functions dynamically by updates to be added to the call and selection rules in case the activator is in the scope of a commit with label l.) Requirement (a) can therefore be formalized by adding to the selection rule the update

Let $path = path(currnode, prun\_pt(l))$
$\forall n \in path$:
  $cands(n) := cands(n) \setminus cands\_1st\_kind(n,l)$
  $\forall m \in children\_1st\_kind(n,l) \setminus path$: $mode(m) := abandoned$

with $path(n, n')$ the set of nodes which lie on the path from n to n', and by adding the following *pruning control condition* on how *currnode* is allowed to be chosen. This restriction is formulated in terms of a mode value *abandoned*:

Pruning control condition: *currnode* never assumes as value a node in mode *abandoned* or below such a node.

Note that if one wants to have both control management outside the system which defines the semantics and a pruning operation inside, then the effect of pruning upon future choices of *currnode* can be formulated only as external condition (as directive for implementations).

In order to satisfy stipulation (b), when a commit $\{Formula\}\_l$ has become empty at *currnode*, one has to delete at $prun\_pt(l)$, where the commit has been introduced, each candidate clause which contains a commit with same label l, and to abandon each child which has been created for executing such a clause. We access those nodes and candidate clauses again by (dynamic and as such easily definable) functions

$$children\_2nd\_kind : \textbf{Node} \times \textbf{Integer} \rightarrow \textbf{Node}^* \cup \{nil\}$$

$$cands\_2nd\_kind : \textbf{Node} \times \textbf{Integer} \rightarrow \textbf{Node}^*$$

Then it suffices to add to the selection rule the following updates [29]:

---

[29] This is a relativization of the *cut* rule for PROLOG in (Börger 1990a; Börger and Rosenzweig 1994a).

$$cands(prun\_pt(l)):= cands(prun\_pt(l)) \setminus cands\_2nd\_kind(prun\_pt(l),l)$$
$$\forall m \in children\_2nd\_kind(prun\_pt(l),l) \setminus path:\ mode(m):= abandoned$$

where *path* is defined as above.

Summing up the preceding discussion, we can formalize the pruning operation by refining the goal update (replacement of *act* by *Bdy* in *goal* to *goal*[*act* ← *Bdy*]), which now includes also deletion of empty commits, and by adding to the previous selection rule the following updates:

let $commgoal \equiv (goal[act \leftarrow Bdy])\ \theta$
$\forall j \in label\_set(Bdy) : prun\_pt(j):= currnode$
$pruning(currnode, commgoal)$

where $pruning(n, goal)$ stands for the updates [30]:

$\forall l \in empty\_commit\_label(goal)$:
Let $path = path(currnode, prun\_pt(l))$
$\quad cands(prun\_pt(l)):= cands(prun\_pt(l)) \setminus cands\_2nd\_kind(prun\_pt(l),l)$
$\quad \forall m \in children\_2nd\_kind(prun\_pt(l),l) \setminus path:\ mode(m):= abandoned$
$\quad \forall n \in path$:
$\quad\quad cands(n):= cands(n) \setminus cands\_1st\_kind(n,l)$
$\quad\quad \forall m \in children\_1st\_kind(n,l) \setminus path:\ mode(m):= abandoned$

The auxiliary function *label_set*: **Clause** → **Integer*** yields the set of all labels occurring in a clause body, *empty_commit_label*: **Goal** → **Integer*** yields the set of all labels l of empty commits $\{\}\_l$ occurring in a given goal.

After the current goal has been updated, in the case of resolution by replacing a literal by the selected clause body, it may contain a commit with empty scope, say $\{\}\_l$. This means that the formula h of a previously introduced commit $\{h\}\_l$ has been computed with success. (Note that by the syntax of GÖDEL, no clause body or initial query has empty commits.) In this case "pruning" must (and by the above pruning update will) be performed[31].

5.3 The computation of negative literals

Computation of negative literals is delayed until they have become ground; this is assured by the abstract *gl_select* function introduced in section 4. If a ground negative literal $\sim lit$ becomes activator, its computation is considered to succeed/fail if the computation of *lit* finitely fails/succeeds.

---

[30] Remember that commits may be nested.

[31] Note that if different commits in different parts of the tree have the same label, then they come from statements in the same procedure definition and have been introduced into the current goal by the same call. This reflects the scoping of labels over a procedure definition and the renaming of labels (together with variables) in calling.

Therefore to compute $\sim lit$, a subcomputation for *lit* has to be started which includes a special action at its exit to report finite failure or success for *lit*. For soundness reasons, pruning has to be disabled inside such subcomputations, which can also be nested. In order to reflect all this in a simple way, we introduce auxiliary marks **neg_beg, neg_end(n)** which during the computation of *lit* act as delimiters of the still to be computed goal [32]. When entering the subcomputation, they are placed around *lit*. We impose that they can become visible to the function *gl_select*, and thereby value of *act*, only as pair bracketing the empty goal, thus signaling that the computation of *lit* has succeeded.

With this proviso, the following *negation start rule* formalizes the call of a negative (ground) literal $\sim lit$: control may now pass to a new node with goal **neg_beg** *lit* **neg_end**(*currnode*). Note that the subtree which has the newly created node as root models the tree associated with the (negation sub-) computation of *lit*. In order to assure that this subcomputation has been exited (by success or failure) before the computation of $\sim lit$ can be terminated, *currnode* has to go into a waiting status, realized by mode *select* with undefined *cands* list and the following *negation control condition*:

Negation control condition: *currnode* never assumes as value a node with negated literal as activator and in mode *select*, unless the child created at this node for the corresponding subcomputation has been *abandoned*.

> **If** *mode* = *call* & *act* = $\sim$ *lit*
> **then Extend** *Node* **by** *t* **with**
>   *mode(t)*:= *call*
>   *goal(t)*:= **neg_beg** *lit* **neg_end**(*currnode*)
>   *sub(t)*:= *s*
> **EndExtend**
> *mode*:= *select*

If *currnode* comes back to a node with activator $\sim lit$ (and therefore in mode *select*), by the negation control condition its only child must have been abandoned and therefore all alternatives for *lit* have produced failure. Then $\sim lit$ is considered to have been computed with *success*. This is expressed by the following *negation success rule*: the computation may continue here in *call* mode, after deleting $\sim lit$ from *goal*. (Note that the computation of a negative ground literal does not modify the substitution attached to *currnode*.)

> **If** *mode* = *select* & *act* = $\sim$ *lit*
> **then** *mode*:= *call*

---

[32] This technique, well known to implementors, has been used with advantage in mathematical form already in (Börger 1990a; Börger and Rosenzweig 1994a).

$$goal := \text{delete } act \text{ from } goal$$

with the obvious meaning of deletion.

The following *negation failure rule* formalizes the case when *lit* succeeds: this provokes failure of $\sim lit$ (easily formalizable, due to pruning control condition, by abandoning the node where $\sim lit$ had to be computed). It happens when *act* becomes neg_beg neg_end(n) in *call* mode, for some node n.

If $mode = call$ & $act = $ neg_beg neg_end(n)
then $mode(n) := abandoned$

To disable pruning inside a negated call (Hill and Lloyd 1992, p.77), it suffices to refine the *pruning* update by putting it under the (easily formalizable) guard:

If *act* is not in the scope of neg-brackets neg_beg neg_end(n) **then**.

5.4 The computation of conditionals

In (Hill and Lloyd 1992) two forms of conditionals IF *Condition* THEN *Formula1* ELSE *Formula2* are distinguished, depending on whether *Condition* and *Formula1* share (local) variables. The version with shared variables, the so-called *Some*-version with guard of form *Some* $[x]$ *Cond* for a sequence x of possibly shared variables, has to be treated separately from what one might call the normal version, because it allows backtracking from *Formula1* to *Cond*. By the *gL select* function definition, *activator* takes a conditional as value only when its guard *Condition* (in the *Some*-version the expression *Some* $[x]$ *Cond*) has no free variables.

Let **X-Conditional** be the set of **X**-conditionals for **X=S**(ome), **N**(ormal). For both cases we apply the marking technique introduced already for negation subcomputations, using delimiters X_cond_beg, X_cond_end(n) with corresponding stipulations for the visibility to *gL select*. The computation of the guard is triggered by the following *guard evaluation rule*, which (in the very same way as the negation start rule) creates a node, decorated in mode call for the computation of the guard, and puts *currnode* into waiting status (realized by assigning mode *select* and undefined *cands* list, given the *conditional-guard control condition* which extends the negation control condition to X-conditionals).

Conditional-guard control condition: *currnode* never assumes as value a node, activated by an X-conditional and in mode *select*, unless the child, created at this node for the computation of the guard, has been *abandoned*.

If $mode = call$ & $act \in $ **X-Conditional**
then Extend *Node* by t with
    $mode(t) := call$

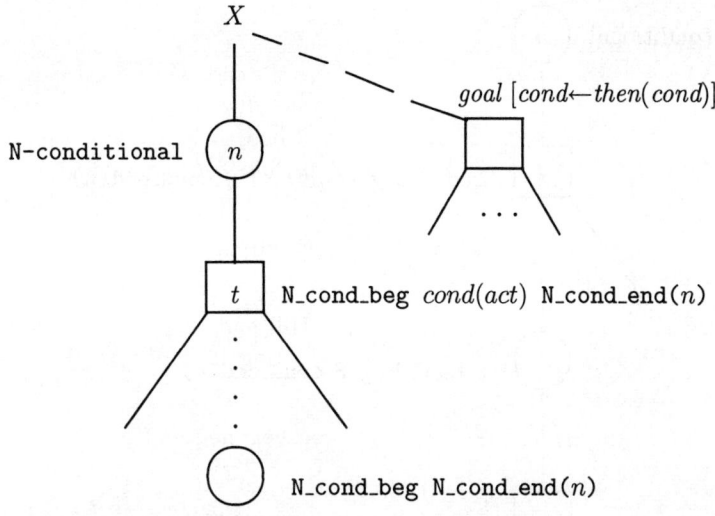

FIG. 1. The computation of an N-conditional's THEN-part.

$goal(t) :=$ X_cond_beg $cond(act)$ X_cond_end$(currnode)$
$sub(t) := s$
EndExtend
$mode := select$

The conditional-guard control condition prevents the demon to come back to the node with X-conditional activator *before* the guard computation started here has been terminated (by failure, see the ELSE rules below, or success, see the THEN rules below).

Due to the stipulations made for *gl_select* wrt the conditional markers, $act =$ X_cond_beg X_cond_end(n) for some n signals that the guard of a conditional has been computed successfully. This triggers the following X_THEN *rules* to create a new node t, ready in mode *call* for the computation of the conditional's THEN-part.

For normal conditionals (X=N) it has to be assured that the system looks only for *one* solution of the guard. By pruning the control condition, it suffices to abandon the node n where the corresponding conditional was called and to create a new node for the computation of the THEN-part[33] (see Figure 1[34]). The goal and the substitution for the new node are defined without applying the substitution which has been computed during the

---
[33] Since *currnode* is below n, we define n and the newly created node to be siblings.
[34] Thanks to Rosario Salamone for production of the two pictures.

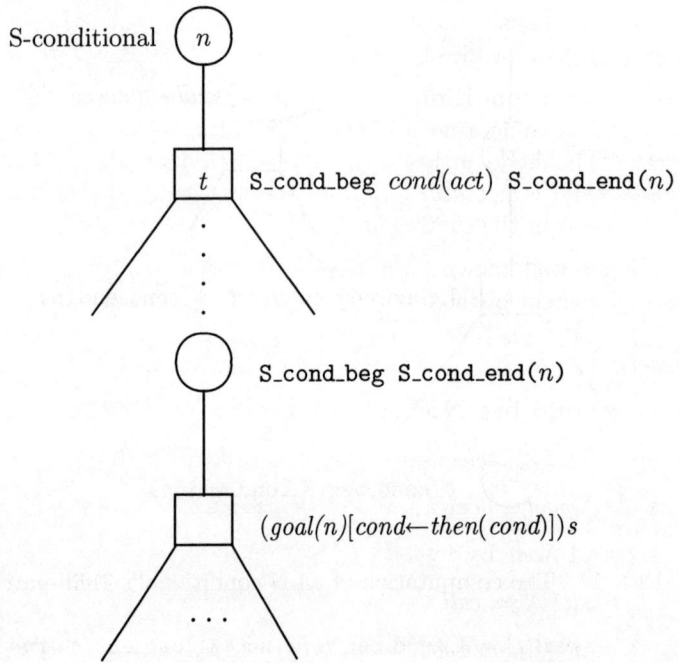

FIG. 2. The computation of an S-conditional's THEN-part.

guard evaluation, reflecting the fact that in a normal conditional the guard (which was ground when called) does not share variables with the THEN-part. (A similar remark applies to the goal update for the THEN-part in the ELSE rule below.)

The corresponding S-THEN rule has similar structure but different updates. Due to the possible presence of shared variables in S-conditionals, the substitution computed by the guard is passed to the computation of the THEN-part and applied to the result of the replacement in defining the new goal[35]. Since from the computation of an S-conditional's THEN-part, the system may (backtrack in an) attempt to resatisfy also the guard, the *mode* of *currnode* is set to *select*[36].

To prevent the system from stepping, by *backtracking* at node n, into the ELSE rules, we set additionally a mark *then_case(n)* to 1. The idea in (Hill and Lloyd 1992) seems to be that once the guard of a conditional

---

[35] For simplicity of exposition we assume here the computed substitutions to be idempotent; otherwise, instead of applying the current substitution s, one should apply only the substitution computed during the just terminated guard evaluation.

[36] Correspondingly the new node is defined to be child of *currnode*. See Figure 2.

is satisfied, computation first proceeds with the THEN-part and only upon failure tries to resatisfy the guard. This sequentiality is reflected in the following restriction of the choice of *currnode*:

S-Then control condition: Let n be an S-conditional node. If *currnode* assumes as value a node in the subtree generated by n, then either there is no node m below n with goal S_cond_beg S_cond_end(n) whose child is not *abandoned* or there is exactly one such node and *currnode* is in the subtree of m[37].

We use the well-known |-notation to exhibit structural similarity but difference in content of the following N- and S-THEN rules:

If $mode = call$ &

$act =$ N_cond_beg N_cond_end(n) | S_cond_beg S_cond_end(n)

then let $cond \equiv gl\_select(goal(n), delaydecl(db), sub(n))$

$newgoal \equiv goal(n)[cond \leftarrow then(cond)]$

Extend *Node* **by** t **with**

$mode(t) := call$

$goal(t) := newgoal \quad | \quad goal(t) := (newgoal)s$

$sub(t) := sub(n) \quad | \quad sub(t) := s$

Endextend

$mode(n) := abandoned \quad | \; then\_case(n) := 1$

$ \quad | \; mode := select$

For S-conditionals, the THEN rule puts *currnode* into mode *select*; if *currnode* comes back to this point (in this mode), it gets abandoned:

If $mode = select$ & $act =$ S_cond_beg S_cond_end(n)

then $mode := abandoned$

If the conditional guard evaluation has failed, this will allow *currnode* to return to the conditional node which originated the guard evaluation subtree and thereby passed to mode *select* (see conditional-guard control condition). In this case, by the following ELSE *rule*, the conditional node will be abandoned and a new node is created [38], ready (in *call* mode and decorated by goal and substitution) for the computation of the ELSE-part of

---

[37] With depth-first search, this apparently complicated condition becomes simple, see the next section.

[38] Due to abandoning of *currnode*, the latter and the new node are defined as siblings.

the conditional. Note that for S-conditionals this is allowed to happen only if the then-case has not been entered; if the THEN-case has been entered, the system cannot enter the ELSE-part and therefore *backtracks* (through the remaining choices for the value of *currnode*).

If $mode = select$ &
  $act \in$ **N-Conditional** | *(act $\in$ **S-Conditional** & then_case $\neq 1$)*
then Extend *Node* **by** t **with**
    $mode(t) := call$
    $goal(t) := goal[act \leftarrow else(act)]$
    $sub(t) := s$
  EndExtend
  $mode(currnode) := abandoned$

If $mode = select$ & $act \in$ **S_Conditional** & $then\_case = 1$
then $mode := abandoned$

(We assume that in conditionals without ELSE-part, *act* is deleted. See also the remark above to the N_THEN rule.)

6 Depth-first search and concurrency

In this section we outline briefly how the preceding definition of semantics of GÖDEL programs can be refined to include depth-first search (as a typical step towards efficient implementation) or concurrency (as the theoretically most challenging extension).

The ground for depth-first search has already been prepared by the definition of the parent children relation used to describe the semantics of pruning. The definition of the *parent* function becomes part of the (semantics defining) evolving algebra by including the corresponding update into the rules. Namely add the update

$$parent(t) := currnode$$

to the rules for selection, negation start, conditional guard evaluation, S-THEN; the update

$$parent(t) := parent(n)$$

to the N-THEN rule, and the update

$$parent(t) := parent(currnode)$$

to the ELSE rules. The depth-first strategy is incorporated into the system by adding to all above-mentioned rules the update

$$currnode := t$$

and setting initially *parent(currnode):= root*. The control conditions are then easily implemented using the backtracking update

$$backtrack \equiv \textbf{if } parent = root$$
$$\textbf{then } stop := -1$$
$$\textbf{else } currnode := parent$$
$$mode(parent) := Select$$

The ϵ-choice condition is implemented by introducing the additional update

If $cands = [\ ]$ **then** *backtrack*

into the THEN-part of the selection rule. The pruning control condition is realized by reinterpreting the update

$$mode(m) := abandoned$$

as $cands(m) := [\ ]$. Under the same interpretation of being *abandoned*, the control conditions for negation, conditional guard and S-THEN are then automatically satisfied. Thus it is a simple exercise to prove the following:

Proposition 6.1 *Let G be the evolving algebra defined in Section 1-5 and G' its refinement as indicated in this section. Then G' correctly implements G under the strategy of depth-first search.*

Now to the question of what is needed to build a parallel implementation of the GÖDEL language. The recent work of Glavan and Rosenzweig (1993) defines a notion of *concurrent runs* within the framework of evolving algebras, which is based upon a notion of *independence* of rules and runs. Using those notions it is easy to extend the above interpretation of the GÖDEL interpreter *I* by allowing different demons to execute concurrently the rules of *I*, restricted only by the independence property. The problem of a parallel implementation of GÖDEL is thereby reduced to the (not at all trivial) problem of implementing an interesting independence notion for our rules (and thereby runs). This corresponds to what has been experienced by us in our formalization of the two forms of parallel logic programming represented by PARLOG (Börger and Riccobene 1993) and CONCURRENT PROLOG (Börger and Riccobene 1992).

7 Conclusion and outlook

The abstract interpreter for GÖDEL programs which has been developed in this paper (refining the core for user-defined predicates by introducing stepwise delay, pruning, negation, conditionals and their implementation under the depth-first strategy) constitutes a mathematical object of manageable complexity. It can therefore serve as a basis for mathematical investigations of computational effects of GÖDEL programs. Our abstract specification of the semantics of GÖDEL programs, which supports the intuitive procedural understanding in a direct manner, can also serve as starting point for a hierarchy of specification refinements leading to actual implementation, where each refinement step comes with a mathematical correctness proof; very much along the lines of what has been achieved for PROLOG with respect to the Warren Abstract Machine (Börger and Rosenzweig 1994b) and for its typed extension PROTOS-L (Beierle et al. 1991) with respect to the Protos Abstract Machine (Beierle and Börger 1992). This may actually provide some methodological help for implementing GÖDEL.

8 Acknowledgements

The authors wish to express their warm thanks to John Lloyd and Pat Hill for careful study and fruitful discussion of many previous versions of this paper. Their criticism and illuminating comments helped us considerably in our attempt to faithfully model their basic intuitions of the GÖDEL programming language.

The work was started when the second author was visiting (March–June 1992) at the Computer Science Department, University of Bristol (UK); it has been partially supported by "Progetto Finalizzato Sistemi Informatici e Calcolo Parallelo" of CNR, under Grant n.90.00671.69.

Bibliography

1. Apt, K. (1990). Introduction to logic programing. *Handbook of Theoretical Computer Science, vol.B* (ed. J.van Leeuwen), 493–574. Elsevier, Amsterdam.
2. Beierle, C. and Börger, E. (1992). Correctness proof for the WAM with types. *CSL'91, 5th Workshop on Computer Science Logic* (eds. E.Börger, H.Kleine Büning, G.Jaeger, M.M.Richter). Lecture Notes in Computer Science vol. 626, Springer-Verlag, Berlin, pp. 15–34.
3. Beierle, C., Meyer, G. and Semle, H. (1991). Extending the Warren Abstract Machine to polymorphic order-sorted resolution. *Logic Programming: Proceedings of the 1991 International Symposium* (eds. V.Saraswat, K.Ueda), pp. 272–288. MIT Press, Cambridge, MA.
4. Blakley, B. (1992). *A Smalltalk Evolving Algebra and its Uses.* Ph. D. Thesis, University of Michigan.
5. Börger, E. (1990a). A logical operational semantics for full PROLOG. Part I: Selection core and control. *CSL'89, 3rd Workshop on Computer Science Logic* (eds. E. Börger, H.Kleine Büning, M.M.Richter), Lecture Notes in Computer Science vol. 440, Springer-Verlag, Berlin, pp. 36–64.
6. Börger, E. (1990b). A logical operational semantics for full PROLOG. Part II: Built-in predicates for database manipulations. *Proc. of Mathematical Foundations of Computer Science* (ed. B.Rovan), Lecture Notes

in Computer Science vol. 452, Springer-Verlag, Berlin, pp. 1–14.
7. Börger, E. (1992). A logical operational semantics of full PROLOG: Part III: Built-in predicates for files, terms, arithmetic and input–output. *Logic from Computer Science* (ed. Y.N.Moschovakis). MSRI Publications vol.21. Springer Verlag, Berlin pp. 17–50.
8. Börger, E. and Riccobene, E. (1992). A Mathematical Model of Concurrent PROLOG. *CSTR-92-15*, Dept. of Computer Science, University of Bristol.
9. Börger, E. and Riccobene, E. (1993). A formal specification of PARLOG. *Semantics of Programming Languages and Model Theory* (eds. M.Droste & Y.Gurevich). Gordon and Breach, pages 1–42.
10. Börger, E. and Rosenzweig, D. (1991a). PROLOG tree algebras. A formal specification of PROLOG. *Proceedings ITI'91 (13th International Conference on Information Technology Interface)*, Zagreb, pp. 513–518.
11. Börger, E. and Rosenzweig, D. (1991b). From PROLOG algebras towards WAM: a mathematical study of implementation. *Computer Science Logic* (eds. E.Börger, H.Kleine Büning, M.M.Richter, W.Schönfeld). Lecture Notes in Computer Science vol. 533, Springer-Verlag, Berlin, pp. 31–66.
12. Börger, E. and Rosenzweig, D. (1991c). WAM algebras: a mathematical study of implementation. Part II. *Logic Programming* (ed. A.Voronkov). Lecture Notes in Computer Science vol. 592, Springer-Verlag, Berlin, pp. 35–54.
13. Börger, E. and Rosenzweig, D. (1994a). A mathematical definition of full PROLOG. *Science of Computer Programming*.
14. Börger, E. and Rosenzweig, D. (1994b). The WAM: definition and compiler correctness. *Logic Programming: Formal Methods and Practical Applications* (eds. C.Beierle & L.Plümer). *Studies in Computer Science and Artificial Intelligence*, North-Holland, Amsterdam.
15. Börger, E. and Schmitt, P. (1991). A formal operational semantics for languages of type PROLOG III. *Computer Science Logic* (eds. E.Börger, H.Kleine Büning, M.Richter, W.Schönfeld). Lecture Notes in Computer Science vol. 533, Springer-Verlag, Berlin, pp. 67–79.
16. Börger, E., López-Fraguas, F.J. and Rodríguez-Artalejo, M. (1994). A model for mathematical analysis of functional logic programs and their implementations. In B. Pehrson and I. Simon (Eds.) *IFIP 13th World Computer Congress 1994, Volume I: Technology/Foundations*, Elsevier, Amsterdam.
17. Glavan, P. and Rosenzweig, D. (1993). Communicating evolving algebras. *Computer Science Logic* (eds. E. Börger, S. Martini, G.Jäger, H.Kleine Büning, M. M. Richter). Lecture Notes in Computer Science vol. 702, Springer-Verlag, Berlin, pp. 182-215.

18. Gottlob. G. and Kappell, G. and Schrefl, M. (1991). Semantics of object-oriented data models the evolving algebra approach. *Next Generation Information System Technology* (eds. J. W. Schmidt & A. A. Stogny). Lecture Notes in Computer Science vol. 504, Springer-Verlag, Berlin, pp. 144–160.
19. Gurevich, Y. (1988). Logic and the challenge of computer science. *Trends in Theoretical Computer Science* (ed. E.Börger). Computer Science Press, Rockville, MA, pp. 1–57.
20. Gurevich, Y. (1991). Evolving algebras. A tutorial introduction. *EATCS Bulletin*, 43.
21. Gurevich, Y. and Huggins, J.K. (1993). The semantics of the C programming language. *Computer Science Logic* (eds. E. Börger, S.Martini, G.Jäger, H.Kleine Büning, M.M.Richter). Lecture Notes in Computer Science vol. 702, Springer-Verlag, Berlin, pp. 274-308.
22. Gurevich, Y. and Morris, J. (1988). Algebraic operational semantics and MODULA-2. *CSL'87, 1st Workshop on Computer Science Logic* (eds. E.Börger, H.Kleine Büning, M.M.Richter). Lecture Notes in Computer Science vol. 329, Springer-Verlag, Berlin, pp. 81–101.
23. Gurevich, Y. and Moss, L. (1990). Algebraic operational semantics and OCCAM. *CSL'89, 3d Workshop on Computer Science Logic* (eds. E.Börger, H.Kleine Büning, M.M.Richter). Lecture Notes in Computer Science vol. 440, Springer-Verlag, Berlin, pp. 176–192.
24. Hill, P.M. and Lloyd, J.W. (1992). The GÖDEL Programming Language. *CSTR-92-27*, University of Bristol.
25. Hill, P.M., Lloyd, J.W. and Shepherdson, J.C. (1990). Properties of a pruning operator. *Journal of Logic Programming*, 1(1): 99–143.
26. Lloyd, J.W. (1987). *Foundations of Logic Programming* (2nd edn). Springer Verlag, Berlin.
27. Lloyd, J.W. (1992). *Private communication*, e-mail of 3 November 1992.
28. Lloyd, J.W. (1993). *Private communication*, e-mail of 5 April 1993.